Product Design - A Manufacturing Perspective

Edited by Evren Yasa and Ozgur Poyraz

Published in London, United Kingdom

Product Design - A Manufacturing Perspective
http://dx.doi.org/10.5772/intechopen.100755
Edited by Evren Yasa and Ozgur Poyraz

Contributors
Ara Jung, Evren Yasa, Farnaz Rezaei, Hüseyin Kürşad Sezer, Kang Hyun Lee, Olcay Ersel Canyurt, Osiris Canciglieri Jr, Oğulcan Eren, Patrick Kenger, Paulo Setti, Stefan Johansson

First published in London, United Kingdom, 2024 by IntechOpen
IntechOpen is the global imprint of INTECHOPEN LIMITED, registered in England and Wales, registration number: 11086078, 167-169 Great Portland Street, London, W1W 5PF, United Kingdom

British Library Cataloguing-in-Publication Data
A catalogue record for this book is available from the British Library

Additional hard and PDF copies can be obtained from orders@intechopen.com

Product Design - A Manufacturing Perspective
Edited by Evren Yasa and Ozgur Poyraz
p. cm.
Print ISBN 978-1-80356-554-5
Online ISBN 978-1-80356-555-2
eBook (PDF) ISBN 978-1-80356-556-9

For EU product safety concerns:
IN TECH d.o.o., Prolaz Marije Krucifikse Kozulić 3, 51000 Rijeka, Croatia,
info@intechopen.com or visit our website at intechopen.com.

Meet the editors

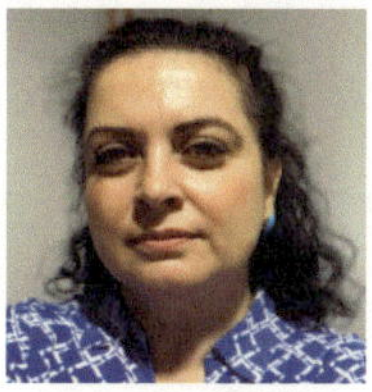

Dr. Evren Yasa leads the Additive Manufacturing group and activities at the Advanced Manufacturing Research Centre (AMRC), University of Sheffield, UK. She graduated with a Bachelor's degree in mechanical engineering from Istanbul Technical University (Turkey) and has actively been working in additive manufacturing since 2005 when she started her Ph.D. studies at the Catholic University of Leuven, Belgium. Her Ph.D. thesis, "Combined Process of Selective Laser Melting and Selective Laser Erosion/ Laser Re-melting," was awarded the Emerald Outstanding Doctoral Research Award. After a short period of postdoctoral study, Dr. Yasa worked as a senior engineer at Tusas Engine Industries, a General Electric joint-venture company specializing in manufacturing aero-engine parts, where she led additive manufacturing projects/activities. Later, she returned to academia as an assistant professor for 5 years. Moreover, she has been working as an independent expert in laser-based manufacturing on behalf of the European Commission. In 2022, she joined AMRC, working closely with the industry to enhance the adoption of additive manufacturing technologies for real-world problems by merging novel research and goal-focused additive manufacturing usability.

Dr. Ozgür Poyraz received his undergraduate and graduate degrees with an emphasis on manufacturing technologies from the Mechanical Engineering Department, Yıldız Technical University, Turkey. He completed the final year of his master's education at Vienna Technical University, Austria, in 2008. He continued his activities in the industry parallel with his academic education, working at Tusas Engine Industries (a joint venture of General Electric Aviation) on different projects, including additive manufacturing of aero-engine components. He completed his Ph.D. in Construction and Manufacturing at the Mechanical Engineering Department, Eskişehir Osmangazi University, Turkey. His Ph.D. thesis focused on modeling the powder bed additive manufacturing of nickel-based superalloys. He worked as a machining research and application engineer on process modeling and simulation of aviation components at Technicut Ltd. Sheffield, UK, for a year. He lectured and supervised undergraduate and graduate students in the Mechanical Engineering Department, Eskişehir Technical University, Turkey. As of May 2023, he started work as an Additive Manufacturing Tech Fellow at the Advanced Manufacturing Research Centre – Northwest. Dr. Poyraz is an author and referee for international journals, books, and conferences. His research interests include additive manufacturing, machining, measurement technologies, process modeling, Computer-Aided Design/Manufacturing/Engineering, and reverse engineering.

Contents

Preface

Product design can be defined as a comprehensive process of imagining, ideating, developing, testing, and iterating products to solve users' problems or to address needs in a specific market. Therefore, it is very critical to identify the market needs, define the problem correctly, and understand the requirements. The designer focuses on the product's functionality, appearance, and manufacturability as well as the experience it will give to its user. Ideally, design for manufacturing (DFM) needs to be integrated into the process very early to ensure the design is optimized and does not necessitate any unnecessary cost due to manufacturing difficulties. This book provides an overview of the product design process and its applications in various fields, particularly manufacturing. Over six chapters, the authors explore such topics as design considerations for additive manufacturing, using modeling approaches to predict the mechanical performance of material extrusion-based AM components, new ways to improve product design, communication-oriented automobile design by a 3D sculpture educational model, and more. Focusing on different engineering and manufacturing problems, this book will help students, engineers, designers, and professionals understand the potential of DFM early in product design.

Dr. Evren Yasa
Eskisehir Osmangazi University,
Eskisehir, Turkey

Advanced Manufacturing Research Centre,
University of Sheffield,
Sheffield, UK

Ozgur Poyraz
Advanced Manufacturing Research Centre NW,
University of Sheffield,
Sheffield, UK

Chapter 1

Introductory Chapter: Overview on Design for Manufacturing and Assembly

Evren Yasa and Ozgur Poyraz

1. Introduction

Product design can be defined as a comprehensive process of imagining, ideating, developing, testing and iterating products to solve users' problems or to address needs in a specific market. The steps of this process are interdependent and should complement each other in order to introduce a successful product. The designers or engineers working on this process have a significant influence on the success of the product. In order to achieve the desired goal, many criteria such as product functions, performance, ergonomics, aesthetics, cost and the launch process must be taken into account by these designers or engineers. With the pandemics, wars and other developments that have blocked logistics routes in the world, especially in the last decade, new criteria have been added, and local sourcing, resilience, agility and production-on-demand subjects started to be considered. While applying many of the criteria listed here contributes to success, ignoring or underestimating them causes the resulting product to have issue(s). In this context, the importance of design for manufacturing and assembly has increased and these guidelines have been implemented by designers and engineers at many steps of the product development process. Moreover, design for manufacturing and assembly rules were included in the concurrent engineering processes, thus saving the total product development time by ensuring that the product development steps were implemented simultaneously instead of a sequential manner. These rules are not only applied in geometric design but are also included as boundary conditions in finite element method, topology optimization and process simulation software. Even though the importance of design for manufacturing and assembly rules has been understood and their application has increased, these rules and related knowledge need to be updated according to emerging manufacturing methods. This book aims to contribute to the required knowledge and, in this context, exhibits the approach to the design of different product groups from a manufacturing perspective.

Design for manufacturing is also known as design for manufacturability and is abbreviated as DFM or DfM. This method simply aims to facilitate the manufacturing, thus reduces the manufacturing times and costs, by changing or refining the product designs. For this process, which is implemented starting from the conceptual design step, product requirements and targeted cost are taken into consideration, and manufacturing methods that can be used to shape the selected product material are investigated [1]. While some products and components can take their final shape with only one manufacturing method, in others it may be necessary to apply more than one method sequentially and/or simultaneously [2]. Regardless of the situation for the related product, as the first step of design for manufacturing, all necessary manufacturing processes must be understood, and their limitations must be revealed [3]. For this appreciation, the designer can refer to the literature on the subject or consult experts in related manufacturing methods. Or alternatively, process modelling software can be utilised, and simulations can be performed.

After understanding the relevant manufacturing method or methods and revealing their limitations, necessary changes are made in the product design. For the current state-of-the-art, design for manufacturing rules is readily available for conventional manufacturing methods like casting, forming, forging, injection moulding, joining and machining [4].

On the other hand, design for manufacturing rules for emerging methods like additive manufacturing is still being developed. While it is still beneficial for design for manufacturing to understand the limitations of additive manufacturing methods, further considerations are needed and improved functionality opportunities should be taken into account such as assembly consolidation, embedded channels/cavities or lightweighting [5]. In addition to these, limitations can be dissimilar to the conventional manufacturing methods and new aspects like minimum wall thickness, minimum hole/cylinder sizes, minimum gap width, the need for support structures and their types and sizes might need to be deliberated [6]. Two different test artefacts are shown in **Figure 1** for the evaluation of hole/boss sizes and stair-stepping effect.

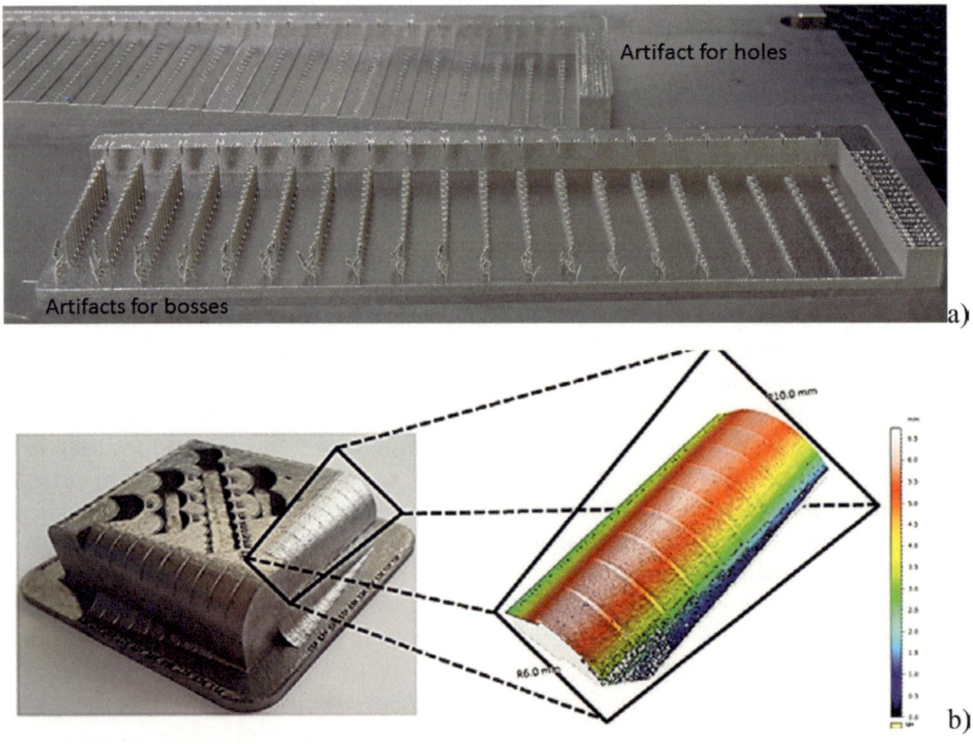

Figure 1.
Design for additive manufacturing test artefacts to understand the limitations: (a) hole and boss test artefacts [7], (b) stair-stepping effect test artefact [8].

DOI: http://dx.doi.org/10.5772/intechopen.114942

In contrast to design for manufacturing, design for assembly rules might be followed to reduce assembly costs and one of the common practices of design for assembly is to make the product integral and combine several sub-components so that the number of assembly operations is reduced. This approach may obviously lead to complex product geometries and increased manufacturing costs. Because of this reason, design for manufacturing and assembly are performed simultaneously and in this way the conflict between two mindsets can be avoided. If the integrated design approach is not followed or not fully adopted, other aspects can be visited to ease the assembling of components. In this respect, decreasing the number of manufacturing operations for each product can be aimed. One of the typical practices to be followed is to use snap-fit designs and to reduce the need for screw/nut type fasteners so that drill or tapping operations are not needed. While this is possible for injection moulded parts, attention should be taken and the reverse angles of snap-fit joints should be design in a way to avoid the need of additional mould cores or inserts.

Focusing on different engineering and manufacturing problems, this book will help students, engineers, designers and professionals to understand the potential of design for manufacturing early in the product design.

Author details

Evren Yasa* and Ozgur Poyraz
Advanced Manufacturing Research Centre NW, University of Sheffield, UK

*Address all correspondence to: e.yasa@amrc.co.uk

References

[1] Handbook ASM. Materials selection and design. ASM International. 1997;**20**:646-670

[2] Farag MM. Materials and Process Selection for Engineering Design. Boca Raton, Florida, United States of America: CRC Press; 2020

[3] Bralla JG. Design for Manufacturability Handbook. 2nd ed. New York, United States of America: McGraw-Hill Handbooks; 1998

[4] El Wakil SD. Processes and Design for Manufacturing. Boca Raton, Florida, United States of America: CRC Press; 2019

[5] Gibson I, Rosen DW, Stucker B, Khorasani M, Rosen D, Stucker B, et al. Additive Manufacturing Technologies. Vol. 17. Cham, Switzerland: Springer; 2021. pp. 160-186

[6] Pei E, Bernard A, Gu D, Klahn C, Monzón M, Petersen M. In: Sun T, editor. Springer Handbook of Additive Manufacturing. Cham, Switzerland: Springer Nature; 2023

[7] Solakoğlu EU, Yilmaz R, Ören S, Akbulut G, Poyraz Ö, Yasa E. Process limits of direct metal laser sintering (DMLS) In terms of minimum hole and boss diameters with varying aspect ratios. In: The 17th International Conference on Machine Design and Production July 12 – July 15 2016, Bursa, Turkiye. 2016

[8] Poyraz Ö, Solakoğlu EU, Ören S, Tüzemen C, Akbulut G. Surface texture and form characterization for powder bed additive manufacturing. Journal of the Faculty of Engineering and Architecture of Gazi University. 2019;**34**(3):1653-1664

Chapter 2

Perspective Chapter: Design Considerations for Additive Manufacturing

Oğulcan Eren, Hüseyin Kürşad Sezer and Olcay Ersel Canyurt

Abstract

The Additive Manufacturing (AM) technology, which was known as rapid prototyping referring to its original usage in prototype production, is progressing fairly well toward full-scale manufacturing of final parts with improvements in precision, strength, and speed. The technology is no wonder a revolution in manufacturing as it allows significant advantages over traditional part making especially for often preferred complex free-form geometries. The AM in theory should allow the production of a simple cube or complex structures of the same volume with the same time and effort. However, this is not always the case; realizing the infinitely design flexible capability of AM with more functional and lighter parts is not a trivial task requiring many iterations and part-specific design considerations such as support structures, part orientations, and interdependent process parameters. Although traditional Design for Manufacturing (DFM) concept has been well documented, Design for Additive Manufacturing (DfAM) is still far from reaching satisfactory levels and correct designs are usually obtained with experience-based heuristic approaches. Here we aim to extract the very best practices in DfAM approaches from open literature and offer a generic guide for engineers designing for AM.

Keywords: design for additive manufacturing, 3D printing, design for manufacturing, design considerations, design to minimize post-processing

1. Introduction

The Additive Manufacturing popularly known as 3D printing is broadly defined by ASTM as a "process of joining materials to make objects from 3D model data" [1]. This definition of rather new manufacturing process essentially refers to a set of technologies in which parts are produced via layer-by-layer material addition as opposed to subtractive methods based on part removal. Hence physical prototypes are easily and cost-effectively produced directly from digital model data without any tooling. The process begins with extracting 3D CAD data in.STL format used in slicer software. The most commonly used. STL (standard triangle language or surface tessellation language) is a file format in which geometries are expressed in finite number of adjoining triangular facets. STL file does not contain any accommodate surface specific data; for example, to identify characteristics such as allowable roughness, intended fit, color, and other surface functions. However, thanks to the user-definable data in binary STL format, relevant surface data can be defined informally [2]. Other data formats such as Polygon File Format (PLY), Additive Manufacturing Format (AMF), and 3D Manufacturing Format (3MF) have been developed to overcome these limitations in STL. The density of these triangles affects the geometrical resolution for curved surfaces (**Figure 1**). For higher resolution, triangulation elements should be denser in complex and curved surfaces, while they can be coarser on planar surfaces.

The process steps of the CAD model preparation for additively manufacturing are presented in **Figure 2**. First of all, the model is initially triangulated depending on the feature complexity described in **Figure 1**. Secondly, the STL model is cut into stack of flat layers by using slicer software (**Figure 2(c)**). These layers describe linear movements (toolpaths) of the extruder, laser or equivalent depending on the additive manufacturing method used. Printing parameters such as layer thickness, printing speed, build orientation, support structures are also determined in the slicer software. At last, since the printing of solid objects requires large amount of materials and time,

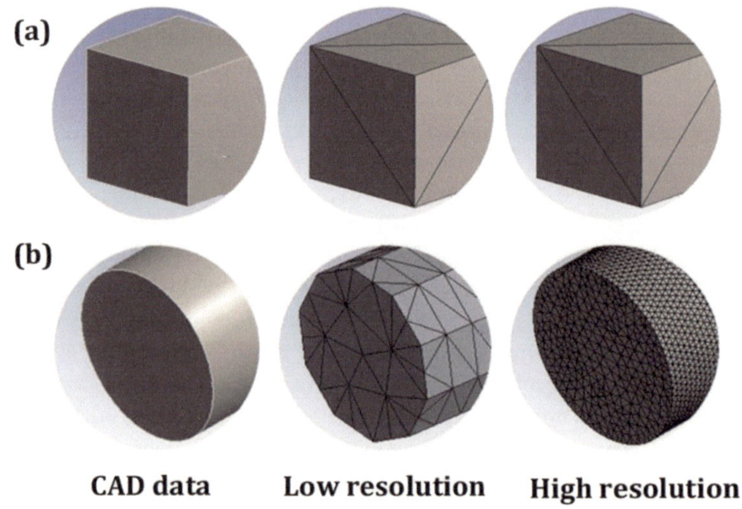

Figure 1.
For higher resolution, triangulation elements should be denser in complex and curved surfaces (a) No need to increase triangular elements, (b) cylindrical part resolution is enhanced with more triangular elements.

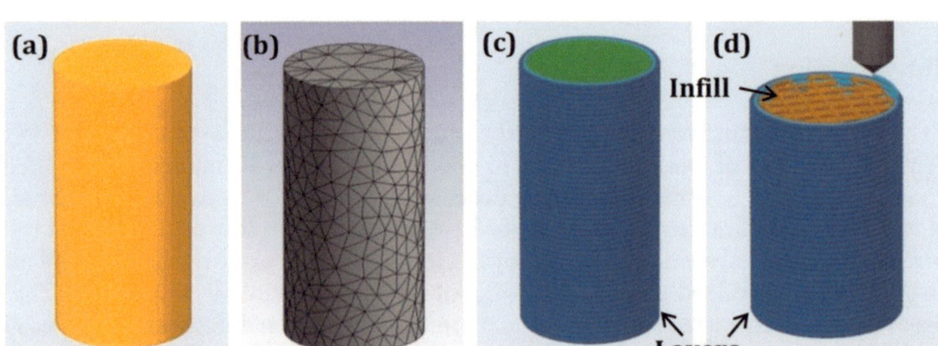

Figure 2.
Additive manufacturing process steps (a). Stl CAD model, (b) triangulation, (c) slicing, (d) print preview showing infill density, pattern, and toolpath.

slicer can automatically convert solid volumes to porous structure with specific infill density (**Figure 2(d)**). These parameters also directly affect the accuracy, quality, and strength of the part [3].

The origin of additive manufacturing is based on the patent on stereolithography in the 1951 [4]. In terms of device, with the patent taken by Charles Hull in the 1980s, the first company to commercialize AM technology with the Stereolithography apparatus was founded [5]. Many different additive manufacturing methods were developed in the following periods. Some of these methods can be listed as follows; Selective Laser Sintering/Melting (SLS/SLM) method in which powder material is laid on a tray and sintered/melted with laser or electron beam, Stereolithography Apparatus (SLA) method in which the photopolymer resin is cured with UV or visible light, or Fused Deposition Modeling (FDM) in which thermoplastic filament is passed through a preheated nozzle and deposited in semi-melted state [6]. Other additive manufacturing methods, layer thickness ranges, materials costs, and usable materials are given in **Figure 3**.

1.1 When to consider AM

Additive manufacturing (AM) methods currently have limited implementation in the manufacturing of final products. However, this is expected to increase with progressive improvements in printing quality, obtaining good surfaces, and decreasing the costs. AM technologies inherently offer design flexibility so that functional, lightweight parts can be produced quickly. The comparison of different production methods in terms of production properties is shown in **Table 1**.

As can be understood from **Table 1**, AM is the most efficient method for the rapid production of specific or assembled parts. While Computer Numerical Control (CNC) machining is used mostly to produce metal parts and Injection Molding (IM) is only used to produce plastic parts, various materials can be used depending on different AM methods as seen **Figure 3**. CNC machining is better than other methods for producing durable parts with high precision. However, when manufacturing with CNC, expert knowledge is needed to determine the tool paths carefully. AM enables

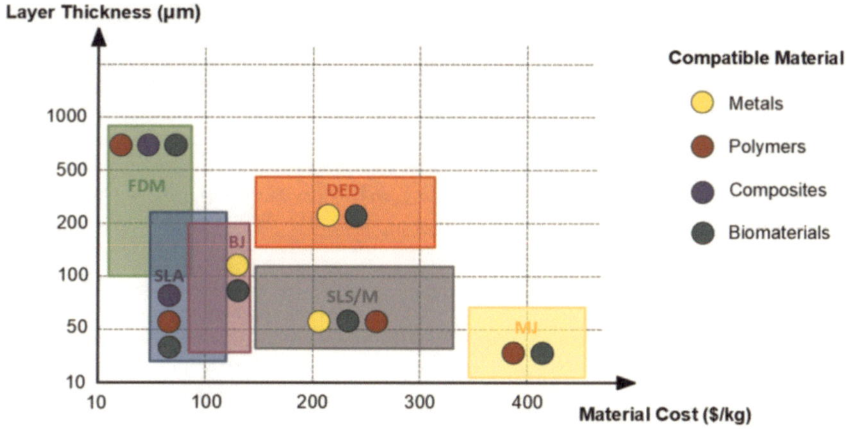

Figure 3.
Layer thickness versus material price chart for AM technologies (selective laser sintering/melting (SLS/SLM), Stereolithography (SLA), fused deposition modeling (FDM), binder jetting (BJ), direct energy deposition (DED), multijet printing (MJ)). The colored circles show the materials that can be used in the related technology [data in figure taken from [7–9]].

Features	AM	CNC Machining	Injection Molding (IM)	Notes
Ability to produce complex parts	∞	✓✓	✓	AM theoretically offers endless design flexibility Mold design in IM depends on strict rules.
Assembled part manufacturability	✓	X	X	AM can produce parts as assembled, No other process can do it.
Mass producibility	✓	✓✓	✓✓✓	IM is the best method for mass production by far. This is one of the weakest point of AM
Light weighting	✓✓✓	✓	X	This is directly linked to ability to produce complex parts. With AM Part topology can be made endlessly complex via topology optimization concept, leaving only material in the design domain. Hence weight can be reduced to its maximum limits in AM.
On demand manufacturing	✓✓✓	✓✓	X	Pre-production in IM is a time consuming and expensive process. Therefore, instead of production on demand, parts are mass produced and stored.
Available material variety	✓✓✓	✓	✓	There are many different AM technologies that are specialized to use different materials. The number of materials that can be used is increasing day by day.
Time between design and production stages	⏱	⏱⏱	⏱⏱⏱	After the design phase in the AM method, the model can be quickly transferred to the printer for production. Mold design and production is required for manufacturing in IM, and tool paths must be programmed before manufacturing in CNC.
Part accuracy	✓	✓✓✓	✓	In CNC, tight tolerances can often be achieved without the need for post-processing. In others, burrs and surface roughness should be removed by post-processing.
Easy to programming	✓✓	✓✓✓	✓	There is no need to create toolpaths in IM.
No need expert knowledge	✓	✓✓✓	✓✓	AM systems are semi-autonomous. Parameters such as printing parameters, support structure and toolpaths can be specified by the slicer software.
Initial investment cost	$$$	$$$	$$$	Mold costs and extrusion system initial costs in IM can reach up to hundreds of thousands of dollars. In AM, it varies from a few hundred dollars to millions of dollars, depending on the printing technology.
No need for post processing	✓	✓✓✓	✓	As the part accuracy is very good in CNC, there is no need for post-processing.
Produced part strength	✓✓	✓✓✓	✓	Porosity may occur in IM and AM.

	Features	AM	CNC Machining	Injection Molding (IM)	Notes
	Low cost per item	✓	✓✓	✓✓✓	In case the number of pieces is over a thousand

Table 1.
The comparison of different production methods in terms of production properties [2, 3, 10, 11].

design-driven manufacturing process, and there is less need for expert knowledge due to limited restrictions. IM is the most preferred method for mass production. Although the initial investment and mold design costs are high, it offers low unit cost per part. However, IM requires very expensive special molds to produce parts with reduced weight, which is not preferred in this specialized method for cost-effective mass production.

More capable and cost-effective printers are being introduced to the market as a result of rapid development of the 3D printing industry [12]. At the same time, there is an increasing demand for special materials that can fulfill the required properties of special parts [13]. However, in order to establish AM methods with high production cost and slow production rate compared to mass production methods, it is necessary to maximize their benefits. The points in which AM can be considered as the most efficient production method to be preferred can be listed as follows [14, 15].

- **Rapid prototyping:** AM speeds up product development by enabling a wide variety of prototypes to be created quickly and inexpensively compared to traditional methods.

- **Customization:** With the design flexibility of AM methods, functional parts tailored to specific needs can be easily produced.

- **Part flexibility:** Almost any geometric form that maintains stability while reducing the object's weight can be designed and produced using AM.

- **Production flexibility:** AM does not require costly and/or time-consuming setup processes to manufacture different products. Thus, enabling more streamlined supply chains with on-demand production.

- **Assembly:** AM allows for the consolidation of an assembly into a single part as opposed to producing numerous pieces and assembling them later.

- **Part reliability:** Strict production tolerances and tightly controlled assembly procedures are often needed for small components, those with complicated elements, or those with small moving parts. AM techniques aid in lowering component defect rates and enhancing part reliability.

- **Material/energy savings:** Compared to conventional manufacturing, AM reduces energy use by using less materials and eliminating steps in the manufacturing process. Additionally, AM uses only as much material as needed to create that part, ensuring minimal waste.

- **Storage and inventory costs:** In AM, components are printed according to demand, thus eliminating the need to keep excess inventory stock and associated transportation costs.

Part manufacturing with AM methods has gained great momentum especially in sectors such as space, aviation, and automotive where part weight and functionality are important [16]. Although AM technologies offer design flexibility, the strength, and surface properties of the part vary according to printing parameters and orientation. Therefore, there are some factors to be considered in the design of the part to be produced by AM manufacturing. This design approach, which takes into account the limitations and unique capabilities of AM, is called DfAM.

2. Design for additive manufacturing

Design for additive manufacturing (DfAM) is the process of product design that uses AM's unique capabilities. DFAM also considers manufacturing constraints and design criteria, which included creativity, applicability, economics, and esthetics of the design [17]. The designer sets the defining and describing the characteristics of the part, known as form, fit, and function referred to as F3 or FFF, design goals, and design validation criteria. Form is the geometric structure of the part. Properties such as shape, size, mass and how the part is seen are evaluated under the form. Fit is the physical relationship of the parts with each other. The positions, assemblies, and tolerances of the parts concerning each other are evaluated within this scope. The function defines the design and intended use of the product. A product can be customized to be used for multiple purposes. So, it is often unnecessary to manufacturing parts designed for traditional manufacturing using the AM method without any changes in their design. Because it will not take advantage of the design flexibility that AM offers, and therefore, the manufacturing cost will likely be more expensive. Thus, to use AM's unique advantages, traditional design must be adapted or redesigned for additive manufacturing [18]. In **Table 2**, comparisons of Replicate with AM, Adapt for AM, and Design for AM belonging to lower wishbone used in cars are given.

In DfAM, mechanical properties are improved by decreasing stress concentrations as well as reducing weight and therefore cost as can be seen in **Table 2**. The designer's strategy to reduce these stress concentrations is based on experience and knowledge. However, computer and artificial-intelligence-based iterative design processes such as topology optimization [20] and generative design [21] used to obtain the right-hand side model in **Table 2** can be helpful methods in DfAM. These complex models, which are impossible to manufacture with traditional methods, can be produced fast, cost-efficient and with advanced mechanical properties using AM's unique production capabilities.

2.1 Fundamental principles of DfAM

While AM seems to offer unlimited design flexibility, there are some factors to consider in the design process for AM. Before starting to design with AM, the suitability of the product for 3D printing should be evaluated. Key factors to consider in the design process for AM by the designer are as follows.

	Replicate with AM	Adapt for AM	Design for AM
Description	No changes are made to the DfM used in traditional methods. The goal is to duplicate an existing product with minimum lead time. This does not affect production cost, speed and part weight or strength	In Adapt for AM, the form is affected. Thus, AM constraints (overhangs, support structures and so on) are eliminated. With Adapt for AM, printing time is shortened and less post-processing is required. But the fit and function criteria of the part are not taken into consideration.	The entire part is redesigned to maximize AM's unique features and minimize its limitations. All Form-fit-function characteristics of the part are defined.
Figure	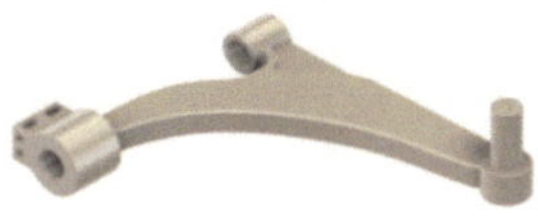	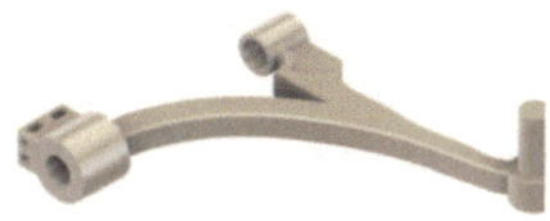	
Safety Factor	1.96	1.75	14
Weight (gr)	4159	2969	2444

Table 2.
Comparison of AM design methods for lower wishbone used in car suspensions [19].

2.1.1 Printing limits

Although the printing volume in 3D printers varies according to the printing method and models, it is generally 16,000 cm^3 (250x250x250 mm) or less. For larger parts, special printers can be used, or modular parts should be printed and assembled afterward. Besides, traditional methods are more cost-effective and faster in the production of large block parts with low complexity. In **Figure 4**, depending on the volume and complexity of the part, additive manufacturing or traditional

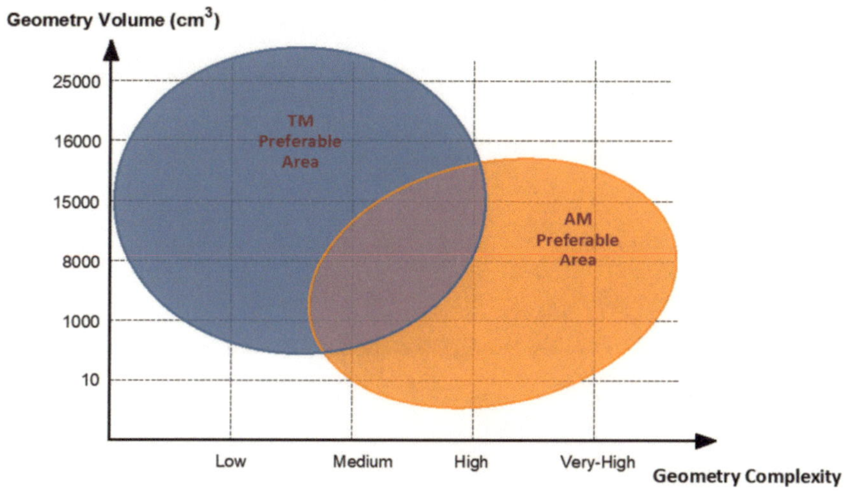

Figure 4.
Additive manufacturing (AM) or traditional manufacturing (TM) regions can be preferred depending on the volume and complexity of the part [adapted from [2, 22, 23]].

manufacturing regions that can be preferred regions are given. The designer should determine the manufacturing method according to the volume and complexity of the existing part. Or the existing design can be updated to take advantage of AM. Thus, low-volume parts can be redesigned and produced easily with AM even if they have complex shape.

2.1.2 Printer parameters

While designing for AM, the method and process parameters to be used should be well known. Because parameters such as laser intensity, layer thicknesses, and scan pattern can determine the limitations or capabilities of the method. In **Figure 5**, the effects of the process parameters on the surface roughness of the AlSi10Mg part produced by the SLM method are given [24]. Here, the surface roughness is negatively affected by the increase in scanning speed, hatch gap, and power, while the increasing energy density increases the surface quality. Besides, the accuracy and quality of the part also depend on the minimum layer thickness and element size. The minimum element size may be related to the printer used, as well as other factors due to the printing method. For example, in the powder bed fusion (PBF) method powder around the part gets partially fused in the part surface, and it affects the surface quality and part tolerance [25, 26]. Hence, holes with high thickness and small diameter may not be printed in PBF [3]. To avoid this situation, the area where the hole is located can be redesigned to require minimum melting or sintering. In addition, small holes with high accuracy can be obtained by changing factors such as layer thickness, printing direction, powder diameter, and laser power.

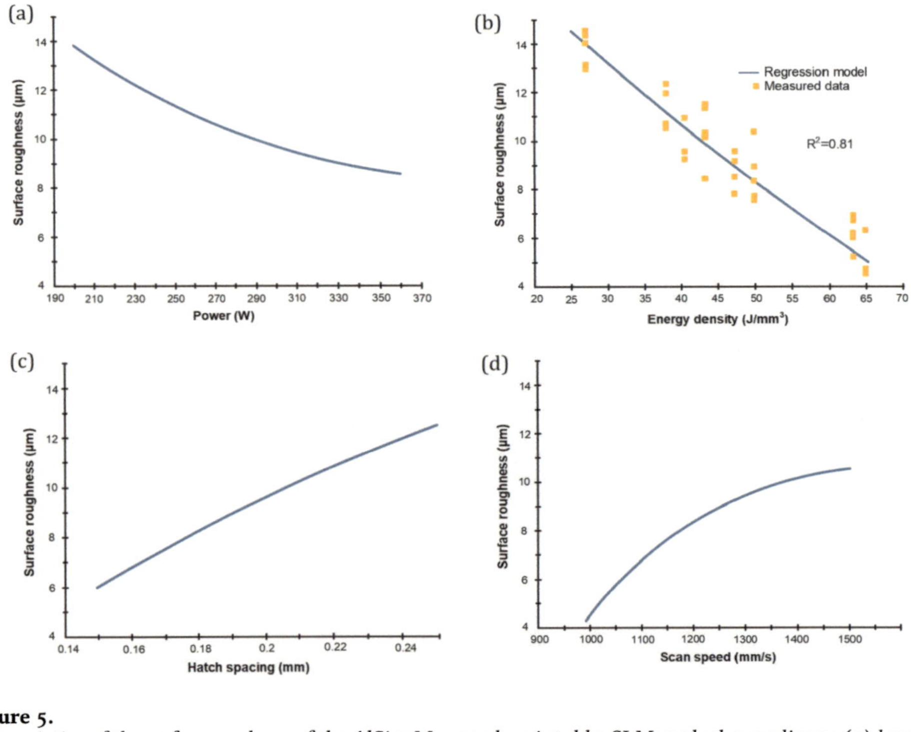

Figure 5.
The variation of the surface roughness of the AlSi10Mg sample printed by SLM method according to (a) laser power (W), (b) energy density (J/mm³), (c) hatch spacing (mm), and (d) scan speed (mm/s) [24].

2.1.3 Affordability

Traditional or other unconventional manufacturing methods can produce coarse parts faster and cost-effectively than AM. For example, considering that the part shown in **Figure 6** does not have red regions, it will be faster and cheaper to produce with 3-axis CNC machining than AM. For this reason, it will be more advantageous to produce parts that do not require changes in design and do not have strategic importance, using traditional methods. In other words, while it is affordable to produce the non-functional reducer cover by casting method, AM is more affordable to produce complex engine block with microfluidic channels that significantly increase efficiency.

Designs for traditional manufacturing (DfM) can be upgraded with the DfAM approach, making them lighter, more functional, and more affordable. Furthermore, depending on the number of the piece to be manufactured, the affordable method may differ. For example, injection mold cost is quite expensive and requires a long lead time. So, it should be preferred if it is desired to produce hundreds of thousands of repetitive parts. The unit price chart for AM and IM depending on the production volume can be seen in **Figure 7**.

The prices of metal AM machine can up to couple of million dollars. These devices are expected to pay off within 2 years. Assuming with a very optimistic estimate that devices are operating for % 80 of the year, which means that the machine produces parts by working non-stop. Because the 20% loss is not caused by the device being in standby mode, but by maintenance, machine cleaning, preparation, preheating time, parts removal stages. In this case, the hourly operating cost of a metal AM device with an initial investment cost of $ 1 million is calculated as $ 71.3 from the formula [3].

$$Machine\ hourly\ \text{operating}\ costs = \frac{\text{initial investment cost} + \text{interest (if any)}}{payback\ period\ x\%running\ time\ x\ yearly\ hours\ (8765\,h)}$$

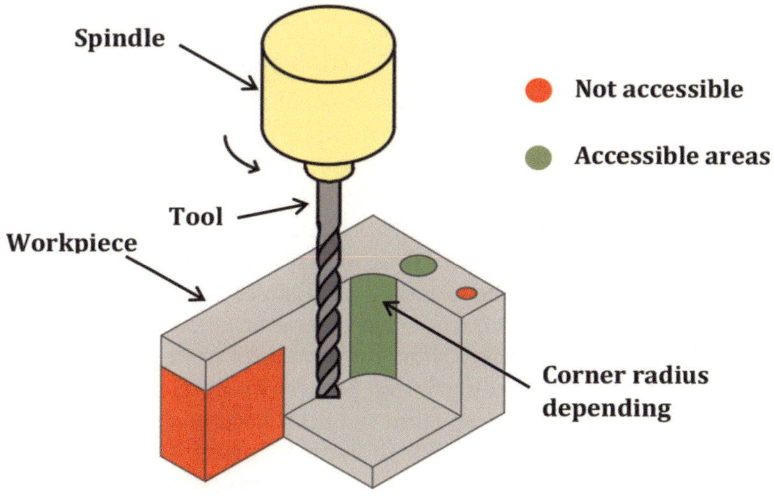

Figure 6.
3-Axis CNC machining schematic. Unreachable areas and holes can be machined, respectively, by increasing the number of axes and using a small diameter tool.

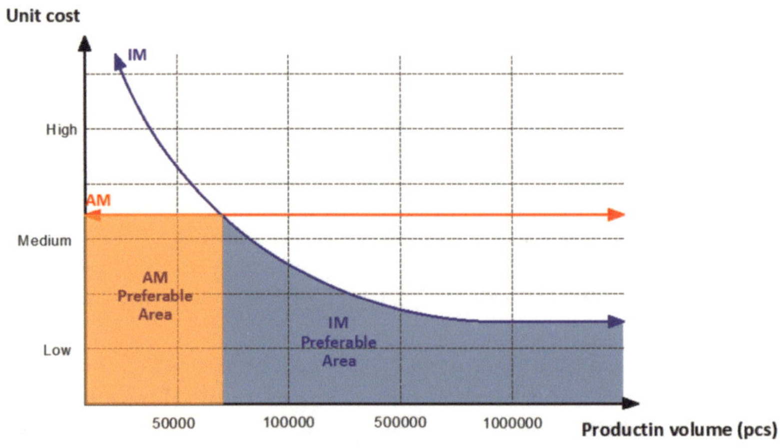

Figure 7.
Unit cost chart for additive manufacturing (AM) and injection molding (IM) depending on production volume. Mold manufacturing costs in IM can be hundreds of thousands of dollars. But then millions of parts can be mass produced. As the number of produced parts increases, the cost per part will decrease as the initial setup cost will be shared. In AM, the volume of production does not affect the cost per unit. Because there is no mold and lead time cost in AM [adapted from [27]].

Material costs are shown in **Figure 3**. Material losses such as powder used in the support structure and partially sintered powder also affect this cost. Hence, with DfAM, the solid volume should be reduced without changing the functionality of the part. Furthermore, designs that do not require a support structure or in which the support structure is a feature of the part are among the most important goals of DfAM. Hourly operating cost also increases in post-processes such as heat treatment, removal of the support structure, and ensuring good surface quality. Moreover, pre-processes such as preparing the printing chamber atmosphere, loading the feedstock, and powder handling also increase the operating cost. According to the 2018 Wohlers report [16], %30–45 of the part costs in AM are due to pre- and post-processing. This means that the hourly operating cost can reach up to $ 100.

2.1.4 Printing time

Although the length of the printing time does not mean the design is good or poor, it is very important to minimize costs. For example, positioning the long part of the piece in the vertical (printing) direction increases the number of layers and thus the number of re-coating. The recoater time varies between 4 and 15 s per layer depending on the metal printer used [3]. This means there may be a 300 min (for 50 µm layer thickness and 10 s recoater time per layer) recoating time difference depending on the positioning of the rectangular prism of 100x10x10 mm lengths. This corresponds to $ 356 in terms of hourly machine operating cost calculated at the above. Besides, the printing chamber must be filled with inert gases (argon, nitrogen, or a vacuum) and heated before the printing process in metal printers. The process of removing oxygen, known as purging, from the print volume and preheating of the build chamber can take several hours. For this reason, printing many parts at the same time and with the long edges on the horizontal axis should be preferred in terms of printing time.

Contrary to traditional methods in DfAM, large masses should be avoided as adding material is costly and time-consuming. Moreover, material accumulation leads to higher residual stresses due to high energy input. One of the frequently used methods to reduce the printing time, residual stresses, and the amount of material used is to create gaps in the part. As seen **Figure 8**, in order to create these gaps, contour lines that will form the outer surface of the part are printed first. Then the inside of the piece is scanned in different patterns and infill rates. As the part has a hollower structure, the amount of material used, printing time, and part weight decrease, but the part strength also decreases. Design strategies that do not affect the strength of the part considerably but maximize the amount of hollow should be developed with DfAM.

It can be used in shell other than hatch pattern to reduce part weight, amount of material used, and printing time. Shelling means that bulk of the material is extracted from interior and only a certain wall thickness remains (**Figure 9**). Thus, the material used and the scanning distance can be reduced by up to 95% [3]. During the shelling process in DfAM, orifice should be added to remove the materials remaining in the interior of part as support.

Reducing print time is one of DfAM's priorities. However, design decisions made to reduce printing time such as part orientation also have a great impact on

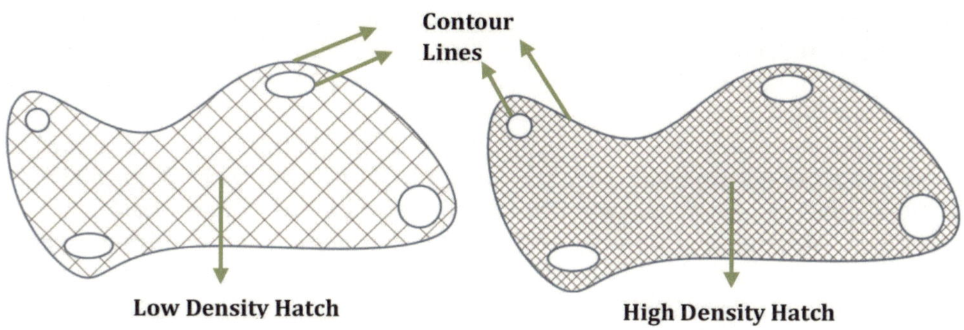

Figure 8.
Printing strategy commonly used in AM methods. First, the outer contour is determined and then the interior is scanned at various infill rates. The low-density hatch pattern provides fast printing of light parts while reducing mechanical properties.

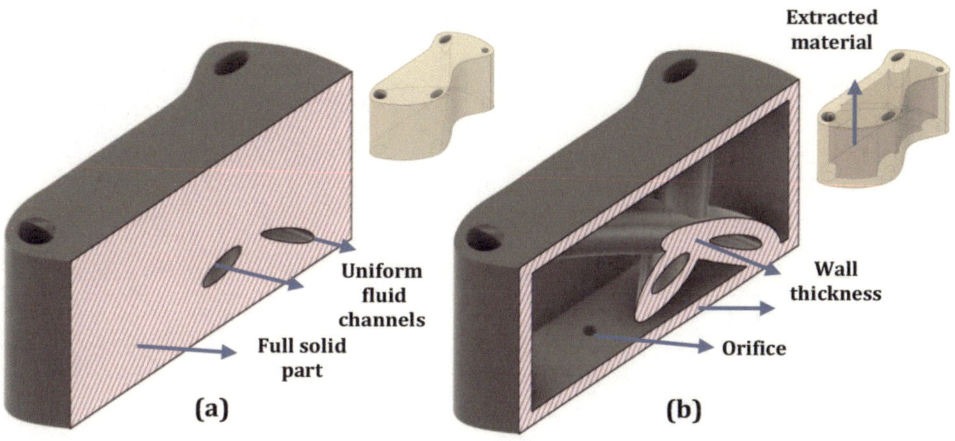

Figure 9.
(a) Solid geometry with complex channel structure, (b) shell geometry with specific wall thickness.

mechanical properties, surface finish, geometry accuracy, and support structure of part. Because of these conflicting effects, the designer must set design priorities. In addition, these contradictions can be minimized with different design methodologies such as TRIZ, which offers possible solutions for these contradictions.

2.1.5 Print orientation

The orientation of the part to the build plate significantly affects material consumption and printing time. Support structures may vary depending on the print orientation. Printing more support structures means extra material, time, and energy consumption (**Table 3**). Additionally, post-process is required to remove the support structure after printing sequences. This increases labor and cost. The support structure not only increases the printing time and cost, but also dramatically reduces the quality of the part surfaces it contacts. For this reason, DfAM aims to design support free parts or positioned on the build plate requiring minimal support.

Another issue that should be considered in printing orientation is holes and cylindrical structures. If the holes and cylindrical structures are printed on the horizontal axis, support structure and stair-step effect will be formed and the geometry will shift from circular to slightly elliptical (**Figure 10**). Besides, the height of the part in the z-axis increases the number of layers. The printing time will be longer as the number of recoating will increase depending on the layer thickness. Moreover, large horizontal surfaces in a single plane result in high residual stresses.

Print orientation affects surface quality regardless of support structure. In the SLM method, as the build angle increases, the surface roughness decreases (**Figure 11**) [25]. However, when the length in the Z direction increases, since more re-coating is required, the printing time and cost will increase. Surface quality can be improved further by changing other factors such as laser power, scanning speed, hatch spacing, laser focus diameter, and shield gases in the environment. In addition to printing parameters, surface quality can also be improved by post-processes such as surface remelting [28], shoot peening electropolishing, chemical polishing, grinding, and sand blasting [29].

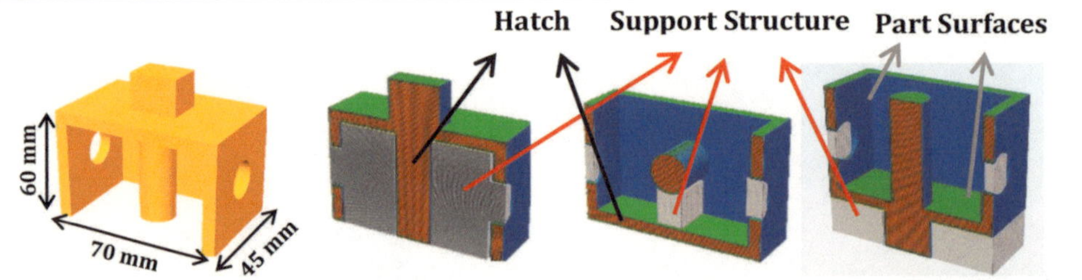

Build Time	18 hours 38 minutes	13 hours 9 minutes	15 hours 21 minutes
Filament length (m)	27.4	19.7	22.9
Plastic Weight (gr)	824	594	688
Material cost ($)	28.85	20.81	24.11

Table 3.
The effect of print orientation on build time, material consumption, and price for the part to be produced with the FDM method.

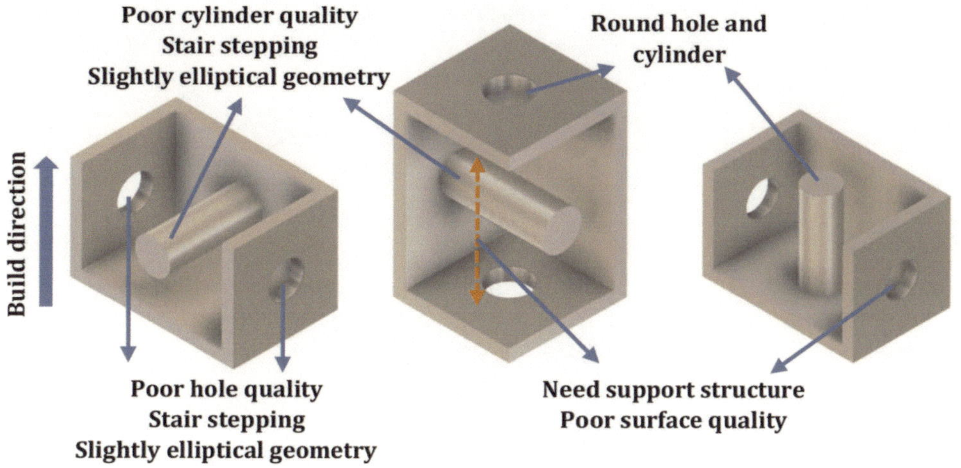

Figure 10.
Hole and cylindrical feature quality change depending on printing orientation. If the circular elements are perpendicular to the printing axis, elliptical structure and stair step are seen. Also, poor tolerance is observed if the support structures contact with the circular elements.

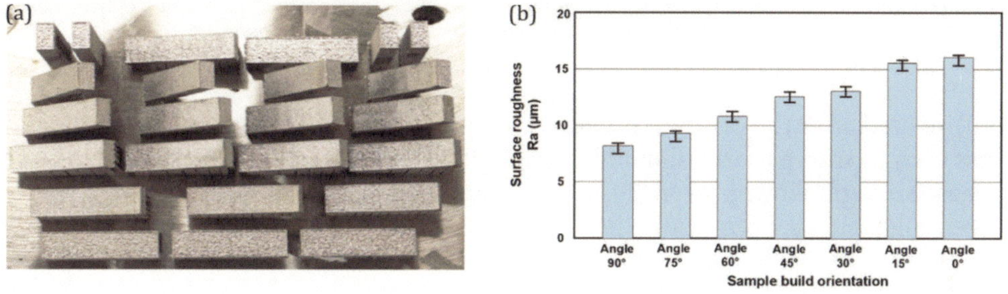

Figure 11.
(a) Building orientations for stainless steel 316 L printed using the SLM method and (b) average roughness (R_a) of inclined surfaces [25].

2.1.6 Part strength

The support structure is not the only factor to be considered when orienting the part to the build plate. The strength, material properties, and surface quality of the part also vary according to the printing orientation. Anisotropy, one of the biggest drawbacks of layered production, occurs in the vertical printing direction and significantly affects the strength. If the part will be exposed to axial loads, the part must be horizontally oriented to stand these loads (**Figure 12**). Although anisotropy is seen in all AM methods, it is most effective in FDM method [30].

While it is largely eliminated in the PBF method with post-processes such as hot isostatic pressing (HIP). Apart from anisotropy, porosity and micro cracks that reduce the strength of the metal printed part can be reduced with the HIP. Also, residual stresses affect the strength of the metal AM parts. Residual stresses arise with the effect of shear forces between layers due to sudden heating and rapid cooling, which is inherent in laser melting. In regions where excess material is fused, there is more residual stress due to the greater thermal input. High residual stress areas are prone to distortion. Therefore, heat treatment such as residual stress relief should be

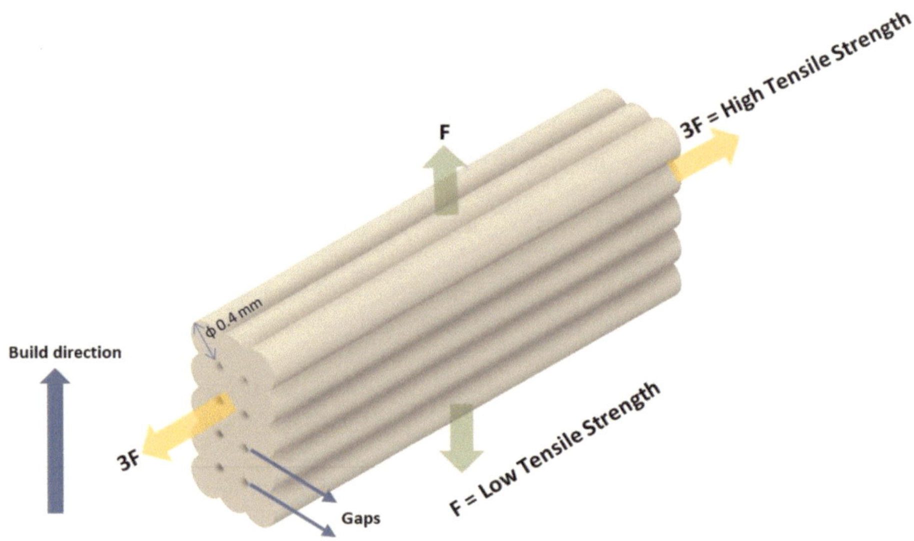

Figure 12.
Anisotropic property for FDM method due to printing orientation. Due to the gaps between the layers formed in the vertical (printing) direction of the part, they are more loosely bonded to each other. Thus, a smaller force causes the breakage. Since there is continuity in the horizontal direction, greater forces can be handled.

implemented after the printing process. Or printing strategies that reduce residual stress such as scan patterns should be specified in DfAM stage. Ribs can be used to increase strength while preventing excess material in a certain area. Thus, while preventing residual stresses and distortions, the amount of material used and printing time are reduced at the same time. **Figure 13** shows the rib structure used to increase the strength of the shell geometry.

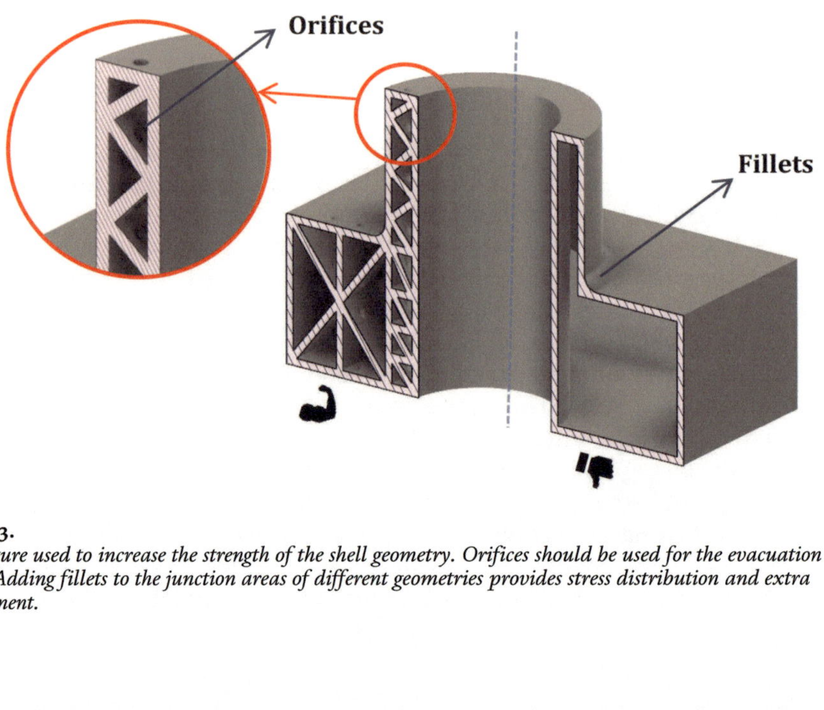

Figure 13.
Rib structure used to increase the strength of the shell geometry. Orifices should be used for the evacuation of excess powder. Adding fillets to the junction areas of different geometries provides stress distribution and extra reinforcement.

3. Design to minimize post-processing

Post-processes may be required for AM to meet features such as microstructure, surface roughness, residual stresses, porosity, and dimensional tolerance. Two types of post processes, thermal and mechanical, are encountered. Heat treatments are performed to improve the microstructure, residual stresses, and porosity of AM parts, while mechanical processes are aimed to enhance the surface properties and dimensional tolerances of the parts. Post-processing steps for products produced with metal AM methods are given in **Figure 14**.

After the printing process, wire electro-discharge machining (EDM) is commonly used to remove the support structures from the product. There is an electrical discharge between the wire and the part, and using heat from electrical sparks, the support material is removed from the work piece by utilizing wire EDM. Then sandblasting, which is based on spraying abrasive particles onto a surface under high pressure, is used to get rid of unfused or partially fused powder remaining on the part after the support structure is removed. Thermal, mechanical, and residual stresses that may cause distortion can occur in the metal AM printed products due to the rapid heating and cooling cycles during the laser melting process. Thermal methods such as stress relieving (preferred in the treatment of thermal and residual stresses) or annealing (preferred in the treatment of mechanical stresses) can be used to remove such stresses. In these processes, the product is kept in a heated (500–800°C) argon or nitrogen atmosphere for several hours and then slowly cooled. A furnace can be used for this process, as well as the printer can perform the same process in its printing volume. Another most used post-process in metal AM methods to enhance the mechanical properties, increase density, and reduce porosity is Hot Isostatic Pressing (HIP). High-pressure argon gas (up to 2000 bar) and high temperature (up to 2500°C) are applied to the AM product in the chamber in the HIP process [31]. Mechanical machining can be used to increase surface quality and add other features such as holes. Surface finishing can also be improved by non-mechanical processing methods such as Chemical Etching and Electro Polishing. One of the primary goals of DfAM is to minimize these post-processes that cause time, labor, and cost. Support structures must be designed for easy removal. There should be no support structure in interior parts that are impossible to remove. Designs that do not need support structures can be created by using DfAM. The circular hole that requires support in its diameter region needs additional post-process such as machining. Instead of circular geometry, water droplet shape makes is possible to remove support structure. If it is

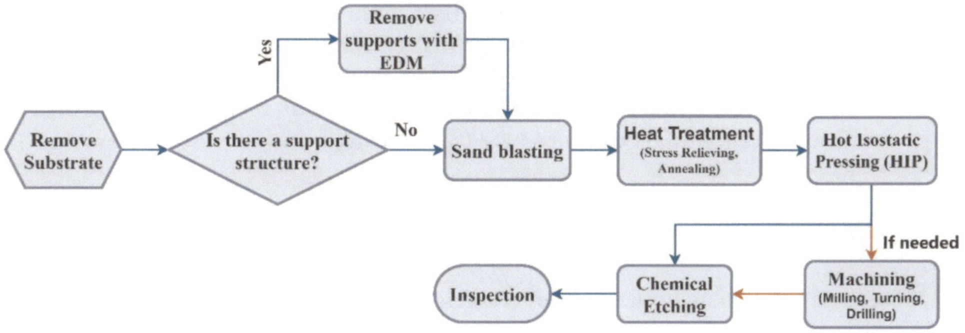

Figure 14.
Standard post-processing steps for metal AM products.

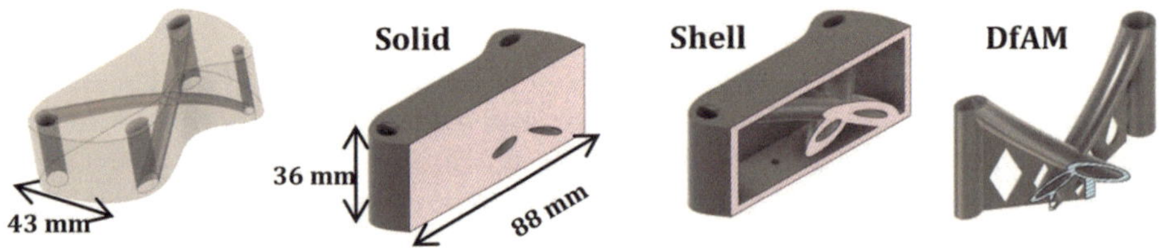

	Solid	Shell	DfAM
Build Time	19 h 6 m	12 h 19 m	3 h 44 m
Material cost ($)	588	418	230
Mass (gr)	225	112	20

Table 4.
Comparison of solid and shell structure based on build time and cost.

unavoidable, the support structure can be designed as a permanent feature that increases the strength of the part. Designing support structures as thin walls also significantly reduces printing time and the amount of material used (**Table 4**).

As can be seen in **Table 4**, compared to the solid model, using a shell structure that may require difficult to remove support geometries significantly reduces parameters such as build time mass and cost. Therefore, support geometries can be designed as a functional element of the part, as shown in **Figure 13**. The same functional part produced following the DfAM methodology can be produced almost five times faster, 11 times lighter, and twice as cheaper than the solid part. Parts with a high strength-to-weight ratio can be obtained using numerical approaches such as Finite Element Analysis. Additive manufacturing of functional structures with reduced weight, such as lattice structures based on the DfAM methodology, also provides improvements in component performance and reductions in production costs and weights [19, 32].

Although it is very effective to design the support structure that causes poor surface quality as a thin wall as a permanent element of the part, thin and large surfaces are prone to warping in the cooling process. Cooling gaps can be created on these surfaces to minimize warping and also to reduce the material used. Or it can be reinforced with other features such as ribs. Post-processes such as machining and chemical etching are required to improve surface quality that contact to support region. The support structure needs to be positioned in the part region where high precision is not required. It is also important to minimize the subsequent heat treatments applied to enhance the properties of the parts such as microstructure, residual stresses, and porosity. Residual stresses and distortions can be minimized by predicting thermal dispersions during the printing process with some specialized simulation software for production with AM. Part-based optimization of printing parameters or laser remelting during the printing process can be performed in order to improve microstructure quality and reduce the porosity without post-processing [33].

4. Conclusion

Overcoming the basic limitations of traditional production methods with AM methods has revealed the necessity of developing unique design innovations for designers. Thus, designers are looking for new direction to customize products, improve product performance, reduce production and assembly costs, and conceptualize products in general. This chapter has covered an understanding of why and when to use AM, an understanding of DfAM methodology, and design considerations. It should not be forgotten that redesigning the product by considering the limitations and capabilities of AM rather than adapting the existing product for AM in the DfAM process yields more effective results in terms of production performance. The factors need to be considered in the design and a number of design examples to maximize AM benefits are discussed for additive manufacturing technologies. There is no doubt about new concepts will be developed for this approach, which is still at the beginning, and traditional CAD and FEA systems will be adapted to evolve the additive manufacturing capabilities.

Acknowledgements

This work was supported by Gazi University Scientific Research Project Coordination Unit [grant numbers: FGA-2022-75320].

Conflict of interest

The authors declare no conflict of interest.

Author details

Oğulcan Eren[1,3]*, Hüseyin Kürşad Sezer[1,3] and Olcay Ersel Canyurt[2,3]

1 Department of Industrial Design Engineering, Gazi University, Ankara, Turkey

2 Department of Mechanical Engineering, Gazi University, Ankara, Turkey

3 Additive Manufacturing Technologies Research and Application Center-EKTAM, Gazi University, Ankara, Turkey

*Address all correspondence to: ogulcaneren@gazi.edu.tr

References

[1] ASTM F2792-12a. Standard Terminology for Additive Manufacturing Technologies. West Conshohocken: ASTM International; 2012

[2] Leary M. Design for Additive Manufacturing. Amsterdam: Elsevier; 2019. ISBN: 09 978-0-12816721-2

[3] Diegel O, Nordin A, Motte D. A Practical Guide to Design for Additive Manufacturing. Singapore: Springer; 2019. ISBN: 978-981-13-8281-9

[4] Munz OJ. Photo-Glyph Recording. USA: Patent US2775758A; 1951

[5] Hull C. Apparatus for Production of Three-Dimensional Objects by Stereolithography. USA: Patent US4575330A; 1984

[6] Mohamed OA, Masood SH, Bhowmik JL. Optimization of fused deposition modeling process parameters: A review of current research and future prospects. Advances in Manufacturing. 2015;**3**(1):42-53. DOI: 10.1007/s40436-014-0097-7

[7] Thomas DS, Gilbert SW. Costs and cost effectiveness of additive manufacturing. NIST Special Publication. 2014;**1176**:12. DOI: 10.6028/NIST.SP.1176

[8] Baumers M, Holweg M. On the economics of additive manufacturing: Experimental findings. Journal of Operations Management. 2019;**65**(8):794-809. DOI: 10.1002/joom.1053

[9] Ding J, Baumers M, Clark EA, Wildman RD. The economics of additive manufacturing: Towards a general cost model including process failure. International Journal of Production Economics. 2021;**237**:108087. DOI: 10.1016/j.ijpe.2021.108087

[10] Renjith SC, Park K, Okudan Kremer GE. A design framework for additive manufacturing: Integration of additive manufacturing capabilities in the early design process. International Journal of Precision Engineering and Manufacturing. 2020;**21**(2):329-345. DOI: 10.1007/s12541-019-00253-3

[11] Altıparmak SC, Xiao B. A market assessment of additive manufacturing potential for the aerospace industry. Journal of Manufacturing Processes. 2021;**68**:728-738. DOI: 10.1016/j.jmapro.2021.05.072

[12] Concli F, Molinaro M, Rampazzo E. Design for additive manufacturing: Is it an effective alternative? Part 2–cost evaluation. WIT Transactions on Engineering Sciences. 2021;**130**:99-110. DOI: 10.2495/CMEM210091

[13] Concli F, Molinaro M, Rampazzo E. Design for additive manufacturing: Is it an effective alternative? Part 1–material characterization and geometrical optimization. WIT Transactions on Engineering Sciences. 2021;**130**:87-97. DOI: 10.2495/CMEM210081

[14] Unleashed. 12 Benefits of Additive Manufacturing and 5 Disadvantages [Internet]. 2021. Available from: https://www.unleashedsoftware.com/blog/12-benefits-of-additive-manufacturing-and-5-disadvantages [Accessed: August 8, 2022]

[15] Attaran M. Additive manufacturing: The most promising technology to alter the supply chain and logistics. Journal of Service Science and Management. 2017;**10**(03):189. DOI: 10.4236/jssm.2017.103017

[16] Wohlers. Additive Manufacturing Report [Internet]. 2018. Available from: https://wohlersassociates.com/press74.html [Accessed: August 8, 2022]

[17] Vaneker T, Bernard A, Moroni G, Gibson I, Zhang Y. Design for additive manufacturing: Framework and methodology. CIRP Annals. 2020;**69**(2):578-599. DOI: 10.1016/j.cirp.2020.05.006

[18] Rosso S, Uriati F, Grigolato L, Meneghello R, Concheri G, Savio G. An optimization workflow in design for additive manufacturing. Applied Sciences. 2021;**11**(6):2572. DOI: 10.3390/app11062572

[19] Eren O, Sezer HK, Yalçın N. Effect of lattice design on mechanical response of PolyJet additively manufactured cellular structures. Journal of Manufacturing Processes. 2022;**75**:1175-1188. DOI: 10.1016/j.jmapro.2022.01.063

[20] Ribeiro TP, Bernardo LF, Andrade JM. Topology optimisation in structural steel design for additive manufacturing. Applied Sciences. 2021;**11**(5):2112. DOI: 10.3390/app11052112

[21] Wang H, Du W, Zhao Y, Wang Y, Hao R, Yang M. Joints for treelike column structures based on generative design and additive manufacturing. Journal of Constructional Steel Research. 2021;**184**:106794. DOI: 10.1016/j.jcsr.2021.106794

[22] Alfaify A, Saleh M, Abdullah FM, Al-Ahmari AM. Design for additive manufacturing: A systematic review. Sustainability. 2020;**12**(19):7936. DOI: 10.3390/su12197936

[23] Taborda LLL, Maury H, Pacheco J. Design for additive manufacturing: A comprehensive review of the tendencies and limitations of methodologies. Rapid Prototyping Journal. 2021;**27**(5):918-966. DOI: 10.1108/RPJ-11-2019-0296

[24] Maamoun AH, Xue YF, Elbestawi MA, Veldhuis SC. Effect of selective laser melting process parameters on the quality of al alloy parts: Powder characterization, density, surface roughness, and dimensional accuracy. Materials. 2018;**11**(12):2343. DOI: 10.3390/ma11122343

[25] Alrbaey K, Wimpenny D, Tosi R, Manning W, Moroz A. On optimization of surface roughness of selective laser melted stainless steel parts: A statistical study. Journal of Materials Engineering and Performance. 2014;**23**(6):2139-2148. DOI: 10.1007/s11665-014-0993-9

[26] Atzeni E, Iuliano L, Minetola P, Salmi A. Redesign and cost estimation of rapid manufactured plastic parts. Rapid Prototyping Journal. 2010;**16**(5):308-317. DOI: 10.1108/13552541011065704

[27] Asnafi N, Rajalampi J, Aspenberg D, Alveflo A. Production tools made by additive manufacturing through laser-based powder bed fusion. BHM Berg- und Hüttenmännische Monatshefte. 2020;**165**(3):125-136. DOI: 10.1007/s00501-020-00961-8

[28] Yasa E, Deckers J, Kruth JP. The investigation of the influence of laser re-melting on density, surface quality and microstructure of selective laser melting parts. Rapid Prototyping Journal. 2011;**17**(5):312-327. DOI: 10.1108/13552541111156450

[29] Löber L, Flache C, Petters R, Kühn U, Eckert J. Comparison of different post processing technologies for SLM generated 316l steel parts. Rapid Prototyping Journal. 2013;**19**:73-179. DOI: 10.1108/13552541311312166

[30] Sezer HK, Eren O. FDM 3D printing of MWCNT re-inforced ABS nano-composite parts with enhanced mechanical and electrical properties. Journal of Manufacturing Processes. 2019;**37**:339-347. DOI: 10.1016/j.jmapro.2018.12.004

[31] Liverani E, Lutey AH, Ascari A, Fortunato A. The effects of hot isostatic pressing (HIP) and solubilization heat treatment on the density, mechanical properties, and microstructure of austenitic stainless steel parts produced by selective laser melting (SLM). The International Journal of Advanced Manufacturing Technology. 2020; **107**(1):109-122. DOI: 10.1007/s00170-020-05072-9

[32] Wang P, Yang F, Li P, Zheng B, Fan H. Design and additive manufacturing of a modified face-centered cubic lattice with enhanced energy absorption capability. Extreme Mechanics Letters. 2021;**47**:101358. DOI: 10.1016/j.eml.2021.101358

[33] Li S, Xin Y, Yu Y, Wang Y. Design for additive manufacturing from a force-flow perspective. Materials & Design. 2021;**204**:109664. DOI: 10.1016/j.matdes.2021.109664

Chapter 3

Applying Successive Wave Iterations to Evaluate Functions and Define Industrial Design Solutions

Paulo H.P. Setti and Osiris Canciglieri Junior

Abstract

The search for competitive edges in the current industry has required conceptual changes in the policy for the development of new economically, environmentally and socially sustainable products. In this direction, assessing the functions of a product and associating them with its manufacturability aspects has been crucial for the cost reduction and for obtaining a more balanced design regarding the value that the customer perceives and the real cost of manufacturing each one of its functions. However, there are still no relevant studies that present the current state of the art regards methods that iteratively interrelate the integrated product development process phases. Firstly, this paper presents a systematic review and content analysis of recent research to define the current frontiers of study. As a second step, it is proposed a model oriented to balance between function value and solutions cost throughout the new products development. Then, to validate, the proposed model was applied to a real case of the consumer goods industry. Among the results, this chapter presents a report showing the relevance of the information collected, the feasibility of the method, its particularities, impacts and limitations.

Keywords: integrated product development process, Project Management, conceptual design, value engineering, concurrent engineering, functions and solutions costs

1. Introduction

In a highly competitive scenario where technology companies need to launch new products more and more quickly, with more quality and with a lower transformation cost, systematic adjustments in the traditional processes of product development (PD) are necessary. Thus, for this search for the continuous evolution of the development process to occur, the scientific world has been working with new proposals and analyzing their effects. This joint effort between academia and the industrial world seeks to develop proposals to complete the formal activities of developing new products within budgetary limits and, above all, generating the possibility of high profitability with the lowest possible risk. For this, each company follows a product development process (PDP) that is more suitable for its business, but which aims to divide the activities of this PDP into phases and stages, in order to systematize the creative process and assist in decision-making. However, whatever the industrial scenario, the definition of phases and stages starts from the project requirements defined by the governing body of organizations based on the voice of the customer and marketing strategies, in order to accommodate the diversity of consumer preferences [1]. These requirements define the restrictions and characteristics of a product and its information must be effectively shared in different phases of the PDP without any loss of meaning [2]. As the requirements are the guiding thread of a product project, all the activities defined throughout this stage of the life cycle keep their eyes pointed at them, but taking into account the needs of all the characters involved in the other stages of life of this new consumer good. Therefore, the interests of the most diverse areas, such as marketing, industrial design, product engineering, manufacturing, quality, logistics, purchasing, technical assistance and the environment, must be discussed at all stages and stages of development. This multidisciplinary work in a concurrent engineering (CE) environment generates collegiate decisions that, at first, reduce the project time by working each step no longer serially, but in parallel, but mainly, reduce future rework.

The need to work in this environment of simultaneous engineering (CE) caused the boundaries between the various departments that existed in technology companies to fall, transforming the product development process into an integrated activity (IPDP). However, even in a simultaneous engineering environment where the various actors involved in the product's life cycle work together at each development stage, it is noticed that some stages are treated very isolated. That is, the walls between the areas of the companies were torn down, but the walls between the phases and stages of the IPDP are still up. As a result, fundamental discussions to define a product, carried out in previous stages, are avoided from returning to the debate and great opportunities can be lost, such as the balance between value and cost [3].

When developing new products, it is important for project teams to understand consumers' perceptions of products, as the success of these products largely depends on the level of customer satisfaction [4]. The search for a balance between value and cost coincides with the search for a project's success, as it has already been confirmed that the better this balance, the greater the probability of commercial success of a new development in the market. As discussed by [5], customer satisfaction is essential for business survival. And this satisfaction is only achieved with the perception of value associated with cost realization. On the other hand, the approaches still treat these two quantities separately, which suggests that the boxes of activities and steps of an IPDP should be open and the discussions no longer take place inside each box, but around them.

Based on this, the objective of this chapter is, firstly, to identify the state of the art regarding the interrelationship and cycles between project phases that discuss value and cost. As well as the frontier of knowledge between the conceptual and preliminary design phases, within which, in a classic way, the perceived values of each function and their associations with the manufacturability aspects generated by the selected solutions are discussed and pondered. From this point, a method is proposed and tested that seeks the balance between these quantities - value and cost - of each function.

For this, this research searched the existing literature for scientific methods or models related to this universe of study. Weighing the relevance, scope and practical applications of the studies and the impact of the respective journals and authors in the area of development of innovative products oriented to technology industries. Thus, this study is initially composed of a systematic literature review, followed by content analysis, with the objective of generating reflections on the interactions between the phases and stages of defining the functions and solutions in the IPDP, culminating in the application of a method in a real industrial design case of the consumer goods industry.

2. Integrated product development process (IPDP)

Integrated product development (IPD) should be understood as a managerial approach to improving product development performance by managing overlap, parallel execution, and concurrent workflow of activities [6]. The study of the steps that make up the integrated product development process (IPDP) in a simultaneous engineering environment shows the importance of using tools to aid in planning and decision-making. As argued by [7], a transfer in the flow of information occurs naturally throughout the life cycle of products and this flow must be systematized. As cost minimization is always one of the goals that a company seeks, the in-depth evaluation of alternatives in a systemic way becomes fundamental to help engineers and other members of a project team to make the decision on the best configuration.

To start an IPDP it is necessary to clearly define which product family will be worked on, which target consumer share and which potential functions can be added. To try help the project team in these issues, there are good methods and development models that use the concurrent engineering premises as a basis and encourage integration between all stakeholders and departments of an organization. Design tools that include value engineering (VE) and design-for-assembly (DFA) techniques are recurrent resources in some of these models, as we will see throughout this chapter. However, the ideal time for application of VE and DFA has not been discussed in depth in the research published so far. Thus, such definitions will be proposed throughout the presentation of this model.

Systematic design methodologies prescribe step-by-step procedures, from functional representation, decomposition and solution search to conceptual, preliminary and detailed design [8]. As argued by [9], due to the increasing number of requirements and the introduction of new technologies, current trends indicate that more disciplines were involved in the design of a product. More and more projects are developed with the participation of actors from the most varied areas, especially in the conceptual and preliminary phases. In order to achieve integration between these multidisciplinary characters, more and more researchers focus on the interface between them, as the information interface plays a very significant role in ensuring that the components defined by designers from different disciplines integrate correctly and eventually help them achieve integrated design.

Always with an eye on performance requirements, which are the constant reverberations that the voice of the customer generates in each decision-making meeting, most authors divide knowledge into two types: explicit knowledge and tacit knowledge – know-how – and the integration of this knowledge of actors from various backgrounds, personal experiences and actions in different departments of a company defines the process of integrated product development [10].

The modern integrated product development process (IPDP) has required simultaneous collaborations of multiple groups, producing and exchanging knowledge and information from multiple perspectives, such as design, manufacturing (including molding, machining, assembly, inspection, etc.), in addition to the disposal, remanufacturing, marketing and other perspectives, within and beyond institutional boundaries [11].

In the same sense, IPDP can be defined as the traditional way of organizing information and defining the steps for creating a new product. Setti et al. [12] show that developing a new project is a complex task that involves all the functional areas of a company and requires the simultaneous work of a multidisciplinary team. This applies both to an isolated product and to a family of products, where one of the important tasks in this complex problem is the determination of optimized architectures of product families, with the product architecture being the arrangement of functional elements that must be foreseen in each one of the formal steps of the IPDP [13].

The steps of an IPDP may vary from company to company, but invariably they are divided into three very well characterized phases, with their own nomenclature. Within each of these phases, design tools are used to assist in decision-making. The project is only "promoted" to the next phase if, in a collegiate manner, each of the intermediate stages is fulfilled. However, what are these phases? As mentioned, the names may vary, but, a priori, they always deal with specific subjects. The first phase addresses issues related to the definition and analysis of the "functions" of the new product. The second deals with issues and decision-making tools regarding the solutions that will be adopted. The third discusses and defines the technical specifications, prototypes and tests.

For some time, as described by [14], it is emphasized that for real integration, the cooperation of teams from various areas of training and interest within a company such as Marketing, Design, Research, Product Engineering, Industrial Engineering, Quality, Logistics, Manufacturing, Sales and After-Sales, for example, that working together and looking towards the end customer throughout all the intermediate activities of the project, define the success or failure of a product. Each design method used in each stage of the IPDP phases must be structured through discussions between the representatives of the areas, forming a simultaneous engineering environment, as illustrated in **Figure 1**.

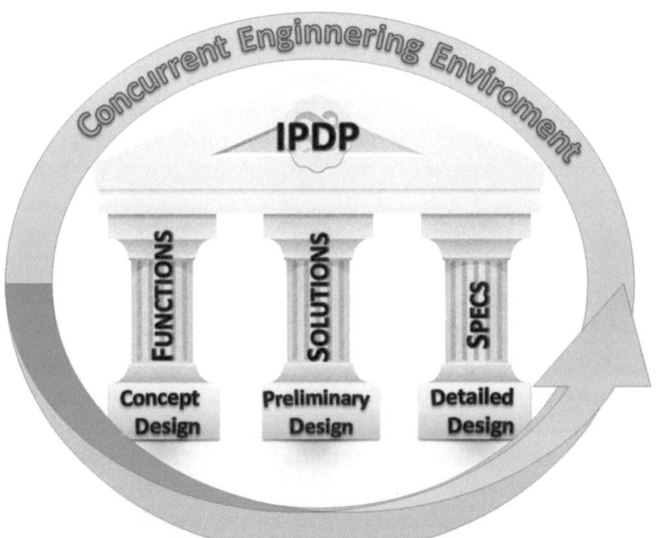

Figure 1.
Three phases of product development immersed in a concurrent engineering environment.

Approaching the development of a new product in this systematic way, the work team naturally ends up considering all aspects related to innovation, production, marketing and after sales. As well as, all the "actors" impacted throughout the life cycle can be listened when was defined the functions and their values, the possible solutions and their costs, and other issues like investment demands and deadlines.

3. Concurrent engineering in the IPDP context

As previously described, a product development (DP) to be considered an integrated development process (IPDP) must structure its activities in a concurrent engineering (CE) environment, where the parallel work of the various members of a multidisciplinary team makes if indispensable. Therefore, the definition of this work philosophy becomes highly relevant.

This concept was first introduced in a report presented by the US Department of Defense (IDA) in 1988 [15, 16]. Since then, this standard of work based on CE has become the main choice for many companies to restructure their processes and carry out an integrated management system, due to significant savings in time and development costs, in addition to improving the quality and quantity of innovations. For the participation of representatives from all sectors of the company in all phases and stages of the IPDP.

In this way, projects, in general, should be understood as a momentary task force. That is, each new product to be developed must be worked on a systemic way with the representatives of the departments that will suffer the direct or indirect effect of it. This development team will be meeting for a predefined time and using all of its technology inventory to include their demands as soon as possible. For this, a matrix management and an integrated work in a concurrent engineering environment becomes essential for any organization to remain competitive and in a constant process of innovation. Therefore, it becomes essential that all levels of an organization understand the philosophy of simultaneous engineering and know how to work in this environment. Each design tool used in each stage of the IPDP phases must be structured through discussions between representatives of the various departments that will be involved throughout the life cycle of that future product, forming an environment of simultaneous engineering, as illustrated in **Figure 1**. With the development of engineering systems, concurrent engineering has received more and more attention.

Recently, many CE technologies have been developed to solve integrated and simultaneous design problems of engineering systems [17]. However, CE can be understood as both a philosophy and the creation of an environment. In the field of philosophy, it is the foundation for meeting stakeholder requirements at the time of conception. Thus, avoiding reworks that generate cost increases and development time. Thinking of the CE as creating an environment, it encourages the project team to work together, with activities in parallel and with the participation of everyone in the decision-making process. This generates gain in opportunities, deadlines and, above all, guarantees the commitment of the agents impacted by the new project.

As described by [18], the process of integrated development of new products is complex, in part due to the broad areas of expertise that need to work together to achieve the best results and from the most varied points of view of representatives from all over the world. The areas involved in the life cycle of a new product that are fundamental to compose the definitive configuration of the functions and solutions of this development and that need to be analyzed. Therefore, each decision must be taken in a collegiate way, that is, with the discussion and active participation of all interested parties, that is, representatives of all key areas of the company that will be part of or will be impacted throughout the entire life cycle of this new product. A well-structured development process can increase the likelihood of success for a new product that would need years of experience from the team members or eventual inspirational insights from an industrial designer, a product engineer or marketer. The systematization of the creative process added to the years of experience and knowledge of IPDP philosophies can be a differential to reduce the probability of failure of a new product in the market [19].

Some researchers argue that this mode of simultaneous work effectively improves the stability of the design process [20]. Collegiate decisions between members from all areas of the company must occur throughout the development process. In this sense, [21] argues that every day, more companies are discovering that the real increase in productivity starts with factors such as clean and efficient processes, good communication infrastructure, simultaneous engineering, collaboration and teamwork. The CE methodology is applied in the early stages of new product development in various sectors of modern industry and is combined with multidisciplinary design optimization. The formalization of the activities of each phase and stage of the IPDP makes the development process faster and the results obtained safer in terms of investment protection, as it seeks to minimize errors and lost opportunities. In addition, it commits all areas of the company in every decision to be made.

The way in which the execution of each activity is defined, according to [22, 23], it should involve each member of this development team to define each aggregate function and each solution defined for the new product. Thus, project development methodologies based on simultaneous engineering tend to have a fundamentally multidisciplinary character conception.

This cycle of product development, which follows the system of first dealing with functions, then discussing solutions and then defining the specifications, in an CE environment, has been treated in the literature as IPDP, where its three formal phases can be described as: conceptual design, preliminary design and detailed design [24, 25]. In this sequence and through its processes, the company is able to create products in a CE environment, more competitive, in less time, being able to meet the constant evolution and demand of the market.

The ultimate target of concurrent engineering is to encourage research and development (R&D), product, manufacturing, PCP, sales and marketing teams to work together in a series of common, pre-established activities. However, people are the ones who make innovation happen, so only with a team trained in the IPDP concepts will the gains be realized. This team must work together from conception. As well as, it must be good communication of decisions made by specialists and by dedicated teams. These stimulator and communicator roles are one of the many activities expected of a good project manager.

The flow of communication and information between specialists is crucial for the success of higher education. In this approach, the conceptual is distributed to all specialists, at which point each one can comment on the project in relation to their own area of expertise [26]. These people who have this expertise must come to a consensus on each decision when conceptualizing the new consumer good. They together, can achieve an optimized product in order to meet the project requirements, add functionality, improve productivity, reduce cost in a shortest possible time. Therefore, design tools to systematize these discussions must be designed and used to organize the decision-making process. At each stage of the IPDP, decision-making must be collegiate, creating an environment of simultaneous engineering between the areas and between the phases of development. With this, the opportunity for innovations and technological synergy is greatly increased through the exchange of information on the heterogeneous competences of the different characters.

4. IPDP phases and steps

Traditional product development (PD) projects are implemented within strict time and resource limits. Technological changes accelerate with intensified global competition, which rapidly changes customer preferences and shortens the life cycles of new products [27]. So, since PD is just one of the inherent activities of the entire lifecycle, it is important to view it within this context. For this, [28] propose that this cycle be divided into seven generic phases that can be adapted to meet the needs of individual projects. They are: Idea generation, Pre-development, feasibility, product development (PD), commissioning, launch and post-implementation review. But they may have alternative names according to the characteristics of the company, as shown in **Table 1**.

Other approaches, such as the one advocated by [29], divide the cycle of a product into just two phases, planning and execution. However, the fine activities are the same and occur in the same sequence listed in the previous discussion and in a continuous cycle between projects. Thus showing that, more important than the name of a given phase, the activities and skills worked on in them are what matters most.

What can be seen is that the most recent studies avoid closing the "boxes", both of the stages of the product life cycle and of the phases and stages of each of these stages, such as the one covered in this chapter, which focuses on the design and development (PD) stage. These same current studies, such as the one presented by [30], in addition to opening these "boxes", propose cyclical evaluations between them, as seen in **Figure 2**. This spiral PDP proposal repeats regular steps, including concept development, system-level design, detailed design, and integration and testing. The process is flexible; the actual number and range of loops may vary.

The PMBOK® guide [31] defines a product design as "a temporary effort undertaken to create a unique product or service". For this, this process must be systematized and organized so that all relevant issues, which will have impacts throughout its

	Product life cycle phases	Alternative names		
Pre-Development	Idea generation	Proposal	Concept	Ideation
	Pre-feasibility	Initial investigation	Initial assessment	Research
	Feasibility	Definition	Business case	Authorization
Development	Development and execution (PD)	Implementation	Construction / Build	Develop and test
	Commissioning	Trial	Beta test	Validation
Pos-Development	Launch	Release	Implementation	Acceptance
	Post Implementation Review	Business review	Project audit	Post project review

Table 1.
The seven stages of the product lifecycle. Adapted from [28].

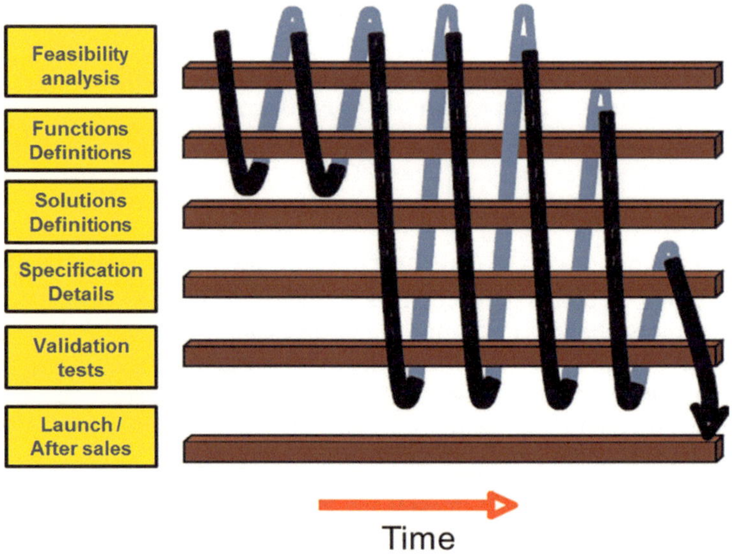

Figure 2.
Spiral between the phases of a product's life cycle. Adapted from [30].

life cycle, are discussed and worked on. The approach of these activities simultaneously between the areas guarantees the improvement of the flow of information and the fulfillment of the project requirements. Thus, the integration of the characters involved throughout the entire life cycle during the development phase generates the concept of integrated product development (IPD).

This chapter argues that there is a strong competitive advantage in the integration of design and development activities, not only among the multidisciplinary agents working together throughout the phases and stages, but also in the integration between the phases themselves. This systematization transforms the DIP into a logical process, where the continuous flow of information would guide the management of activities and facilitate decision-making. We call this, then, the integrated product development process (IPDP), within which three phases with very well-defined activities and distinct focuses are listed. The first one where the functions to be delivered to the final consumer are discussed, evaluated and defined, the second where solutions that meet all project requirements are discussed and defined and the third where the specifications, simulations and tests are elaborated.

Therefore, in an extended planning horizon, with well-defined stages within each phase, more simultaneity in activities is allowed, making decisions collegiate. However, the checkpoint concepts that are applied in many companies can make the decision-making process purely linear. This extreme linearity ends up resulting in missed opportunities and limited ability to adapt to the inevitable changes in scenarios that occur in any type of development. Thus, as far as possible, the flow of information about decisions should occur in both directions of the IPDP [32].

Thus, more than a name for each stage, phase and stage, this chapter aims to discuss the content of this information and reflect on the flow of this information within the IPDP. More specifically, when issues related to the added value of functions are discussed during design and when solutions that generate specific costs for their industrialization are defined.

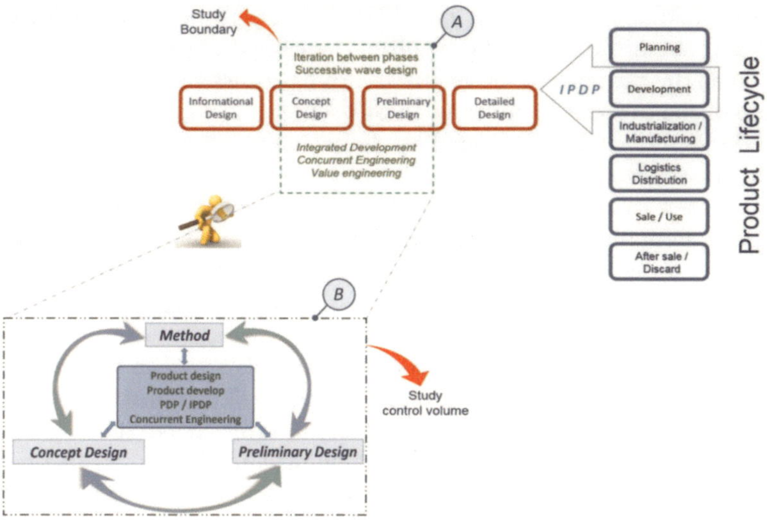

Figure 3.
Defining the location of research in a product's life cycle.

The chapter basis are the activities that happen at the boundary between conceptual design and preliminary design. Throughout these stages, links are structured between the value evaluation of the added functions and the cost generated by the choice of solutions. **Figure 3** shows the limits of this study. Throughout the great journey that defines the product lifecycle, the study focuses on the development phase. Within the development, the frontier of the study is highlighted in detail "A", which involves the phases of definition of functions and solutions that are made during the conceptual and preliminary project. Detail "B" of the **Figure 3** represents the control volume in which the method proposed in this article is found. Where iterations are proposed between the conceptual and preliminary design phases, aiming to balance value and cost in an integrated and simultaneous way in an engineering environment.

To summarize, in the IPDP, the starting point is always a requirements analysis. Afterwards, this analysis is used in a conceptual design, that is, to develop a concept that meets the requirements through functions. From this moment on, the selected concept is designed in order to create maps of possible solutions throughout the preliminary design. After the best combination and solutions are selected, the design is detailed.

The method proposed in this chapter is located at the frontier of the functions definition and solutions definition phases of a new develop. For this, it uses value engineering and cost analysis as a link between these phases. Thus, these concepts will be presented below for a better understanding of the method.

5. Proposal for a model oriented to the balance between value and cost - DFB

The generation of alternative solutions for the same functions is a central element in product design. The search for solutions using morphological matrices explores and encourages the multidisciplinary development group to identify new combinations of principles for existing solutions. Ma et al. [33] argues that the use of the morphological matrix can help the project team to organize the process of generating and combining solutions. Thus, the rationalization of the creative process at this stage lies in the organization of the proposed solutions. In this sense, and supported by a progressive, iterative and integrated ideal defended by the main authors of the systematic literature review, the traditional concept of DFA can be considered as a philosophy of the preliminary project that will serve as a link for the consolidation of the proposed method. This systematization proposal, philosophically, will guide stakeholders to define an ideal number of parts and assembly combinations that optimize material expenditure and the future production process. Always trying to balance the value of the functions with the costs of the solutions.

Bibliographic analyzes showed that the cyclical tools between the functions definition phases and the solutions definition phases are still in an intermediate stage of development [13], making the proposal addressed in this research of significant importance in the initial stages of creating a new product. Thus, [34] proposed an integrated product development method based on CE and DFA, called DFB, from the English acronym "Design for Balance". This method suggests carrying out iterations between the development phase that studies the value of functions (traditionally called conceptual design) and the design phase that studies the solutions for each function (traditionally called preliminary design). If the sum of the modules of the differences between values and costs of all functions is equal to zero, we would have an ideal project. Eq. (1) proposes this goal of perfection. In it, "V" represents the percentage value of each analyzed function and "C" represents the percentage cost to produce each of these functions. That would be a utopianly perfect project. Thus, the proposal of the method is to seek combinations of solutions that approximate, as much as possible, the result of this equation to zero. After all, when the sum of the difference modules was equal to zero, the perfect balance between value and cost of all functions in the studied set would be reached.

$$\sum_{i=1}^{n} |V_i - C_i| \to 0 \tag{1}$$

To achieve this approach, the method proposes iterations between the design phase that studies the value of functions (traditionally called conceptual design) and the design phase that studies solutions for each function (traditionally called preliminary design). In an ideal project, the sum of the differences in values and costs of all functions, in modulus, should be zero. The equation below illustrates this objective, where V is the percentage value of each function and C the percentage cost of each function. As this is utopian, the objective of the method is to bring the result of this equation as close as possible to zero, as proposed in eq. (1).

To carry out the evaluation of the method, [35] presented and tested a proposed method with the five steps described below, as summarized conceptually in **Figure 4**.

The method's intention is to organize the decision-making process, so that ideas can go beyond the involuntary stimuli throughout the developmental routine and become stimulated throughout the application of the method in a systematized way. The proposal is composed of five steps, as step by step below:

Step 1 - Definition of mechanical assemblies: The assemblies must be chosen in view of the greater potential of cost reduction in raw material, without reducing quality, as well as those that present many assembly operations or logistical movements during the industrialization process.

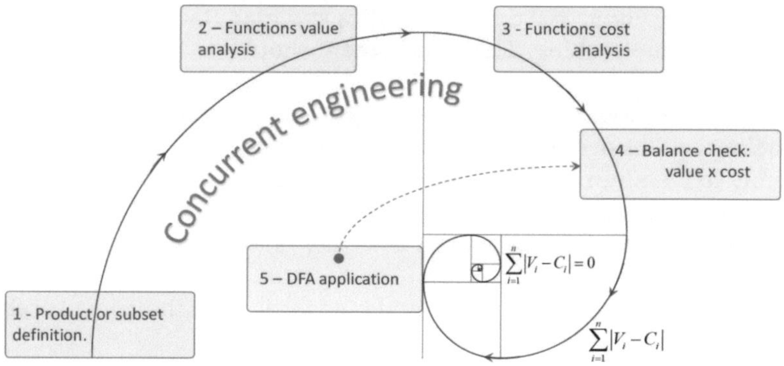

Figure 4.
Concept-based on Fibonacci spiral [3, 35].

Step 2 - Definition of value engineering strategies: The analysis on the use of VE tools will have the purpose of identifying the recommendations for the improvement of solutions that aim at balancing the functions listed in the value x cost chart. The "value" data, which are the percentage levels of importance perceived by the consuming public, will come from the Mudge diagram. On the other hand, the data on the "costs" of the same functions will come from the matrix of cost reduction.

Step 3 - Functions cost analysis: At this stage, the cost analysis of the functions is carried out by the multidisciplinary design team that should discuss the costs of each function and the application of DFA methods in product development.

The analysis of the results shows that the function determined as the main function in the previous step has a higher percentage of cost since it is perceived by the consumer as the most relevant and concentrates the development team's improvement efforts. This cost-rate matrix allows to visualize and understand the cost of each function and the possible impact on product development and reflecting the product success.

Step 4 – Balance check: value x cost: The reduction of any uncertainties in the processes of creation and selection of concepts and functions is necessary to map the boundary conditions associated with the design problem. In this Step, a new table is created, and the functions listed on the horizontal axis and the cost percentages are presented on the vertical axis. Thus, the percentage value derived from the Mudge diagram calculated in Step 2 and the percentage cost derived from the cost allocation calculated in Step 3 can be compared function by function.

The comparison results also highlight which functions have the greatest potential for improvement in terms of equilibrium "value" x "cost". It is worth noting that if an isolated function has a percentage cost higher than its importance, other functions will pay this percentage. Next, the model equation is applied to the difference between the functions and cost, resulting in a percentage showing this relationship's state within the product design. This result reveals the design balance level, and the more the result tends to zero, the more balanced is the design.

The sum of all difference modules between the value and cost percentages provides an indicator that compares the new solution combinations proposals. Therefore, the multidisciplinary product development team's actions will focus on optimizing the solutions combinations proposals and the balance of the design.

Step 5 – DFA application: For the practical application of the DFA concepts, it will initially be necessary to create a multidisciplinary team with representatives of product engineering, industrial design, logistics and manufacturing, who can work together in all decision making. This requirement should be met through in-person meetings.

The DFA methods aim to improve the project, looking for shorter times and costs; thus, this activity will use methods that allow the number of components involved in each assembly to be minimized, and verifying the real need of each one. As a case, the Boothroyd-Dewhurst method (Boothroyd, 2001) was used, in which three simple questions are proposed verifying the real necessity of a component:

1. Does the piece, component or part under analysis have movements related to its surroundings?

2. Does the piece, component or part under analysis require a different material for its function/functionality?

3. Does the piece, component or part under analysis need to be disassembled/ removed for the repair of another one?

If all the answers are "NO", we can propose the unification of components and materials for reducing the cost of raw material and labor.

This step's objective is to verify the most viable configuration in terms of the balance between the "value" and the "cost" of the functions and the lower final cost, considering all the industrial impacts of the proposal in the context of DFA concepts application. The multidisciplinary team analyses the previous step results based on the principle that the design is balanced when the sum of all difference modules between the functions value and cost percentages is equal or close to zero. Next, using DFA concepts proposes a new solution with fewer parts and assemblies, leading to a new material and labor costs chart and, consequently, a new cost-per-function rate matrix.

At this point, the model proposes a further verification of the potential opportunities for improvement and innovation of solutions so that a new product could perform its activities at a lower cost and with the largest scale of perceived value. The model indicates the repetition of steps 4 and 5 at least once, but with as many repetitions as necessary, according to the product's design team definitions and objectives.

The proposed model aimed to encourage the project team to work in an integrated way, systematizing the creative process. With this, increasing the chances of including new opportunities throughout the conceptual and preliminary development and seeking innovations in solutions, with lower cost, presenting the highest possible perceived value and in the shortest possible time. These steps were applied in home appliance industry cases, where excellent results were obtained. The results were positive both in the search for the balance between value and cost of the functions, as well as in the general reduction of material and labor costs of the studied subassemblies.

6. Application case of the model oriented to the balance between value and cost - DFB

In search of an exploratory study to verify the effectiveness of the method, this chapter used as a reference the optimization of the assembly time and the total cost of the solution for the hinge of an appliance to be produced on a large scale.

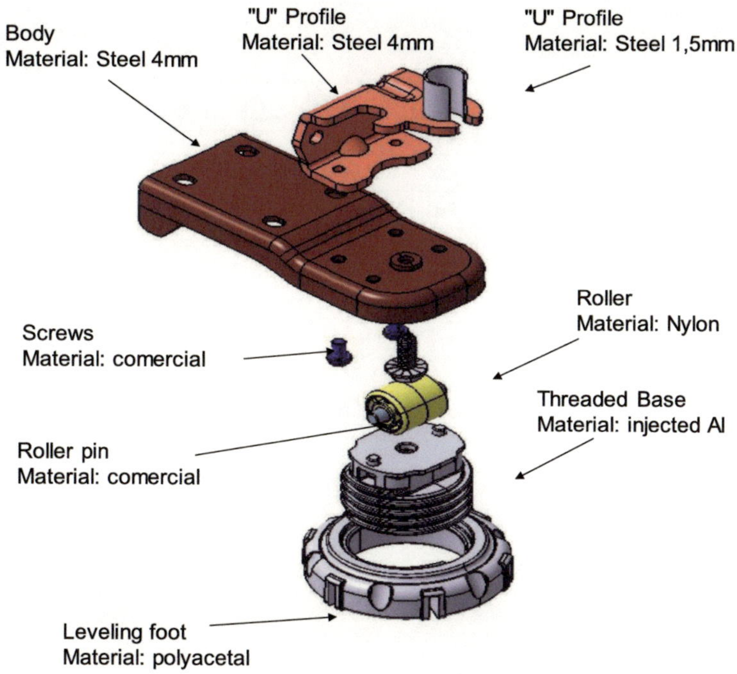

Figure 5.
Home appliance hinge – Original solution.

Figure 5 shows the parts involved in the original solution for this subset, as well as their respective costs of materials and transformation. For the set of solutions contained in the original production condition, a functional analysis was prepared, where the following functions of this set were listed:

Function A – Allow assembly with the product lying down;
Function B – Position and support the door;
Function C – Structuring Cabinet;
Function D – Level the product;
Function E – Slide the product easily;
Function F – Lock the product in the position of use;
Function G – Do not tip the product when the door is opened;
Function H – Enable door replacement for maintenance.

After defining the functions, as indicated by [35], the Mudge method was applied to analyze the value of the functions in terms of percentage importance. Which resulted in the diagram contained in **Figure 6**.

With the functions defined, we started to list and quote the parts, as well as the cost of assembly or transformation (DL) of the original version, as indicated in **Table 2**.

With the parts list formalized and with the respective quotations, it was possible to apply the cost apportionment method to analyze the absolute and percentage cost of each function, through the transposition of the costs of the parts in costs of functions. For this, we used the cost apportionment method described by [35], which resulted in the apportionment matrix shown in **Figure 7**.

With the results of the Mudge Diagram and the Cost Sharing Matrix, it is possible to define the Value x Cost comparison chart of the original product concept, shown in **Figure 8**. To build this chart, the functions were listed on the horizontal axis and the

A	B	C	D	E	F	G	H	Points	Relative importance (%)
A	B3	C5	D3	E3	F1	G3	A1	1	**1,3**
	B	C5	D3	E1	F1	G1	B3	6	**7,9**
		C	C5	C5	C5	C5	C5	35	**46,1**
			D	D1	D3	G1	D3	13	**17,1**
				E	E1	G1	E1	6	**7,9**
					F	G3	F1	3	**3,9**
						G	G3	12	**15,8**
							H	0	**0,0**
						TOTAL		76	100

Figure 6.
Mudge diagram applied to the hinge assembly.

PARTS		COSTS (R$)
	Direct Labor Cost (DL)	5.55
1	Body - Material: 4 mm steel	5.15
2	"U" profile - Material: 4 mm steel	3.30
3	"U" profile - Material: 1,5 mm steel	0.95
4	Threaded base - Material: injected Al	4.60
5	Roller - Material: Nylon	2.55
6	Leveling foot (Sneaker) - Material: Polyacetal	3.60
7	Roller pin - Material: commercial	0.55
8	Screws - Material: commercial	0.20
	TOTAL (R$)	**26.45**

Table 2.
Original assembly parts list and quotation.

A		B	C							
			FUNTIONS							
PARTS		COSTS (R$)	A	B	C	D	E	F	G	H
	Direct Labor Cost (DL)	5,55	12,5%	12,5%	12,5%	12,5%	12,5%	12,5%	12,5%	12,5%
1	Body - Material: 4mm steel	5,15	10%	10%	40%	10%			20%	10%
2	"U" profile - Material: 4mm steel	3,30		25%	35%				25%	15%
3	"U" profile - Material: 1,5mm steel	0,95	20%	60%					10%	10%
4	Threaded base - Material: injected Al	4,60	5%		15%	35%	25%	20%		
5	Roller - Material: Nylon	2,55			20%	10%	60%	10%		
6	Leveling foot (Sneaker) - Material: Polyacetal	3,60			15%	40%	5%	35%	5%	
7	Roller pin - Material: commercial	0,55					100%			
8	Screws - Material: commercial	0,20	20%	10%	30%	10%		10%		20%
	TOTAL (R$)	26,45	1,67	2,62	5,71	4,53	4,10	3,15	2,82	1,84
	Functions costs (%)	100%	**6,3%**	**9,9%**	**21,6%**	**17,1%**	**15,5%**	**11,9%**	**10,7%**	**7,0%**

Figure 7.
Cost apportionment of the original version.

percentages, both of value and of costs, were presented on the vertical axis. Thus, the percentage value, derived from the Mudge diagram, and the percentage cost, derived from the cost breakdown, can be compared function by function.

It can be seen, then, that with this combination of original solutions, there is a significant distance between value and costs in some functions, such as C, D, E, F, G and H. Adding the modules of the differences between value and cost of all functions, we arrive at 59.2. %, as shown in **Figures** 7 and **9**.

The information in this diagram made it possible to trigger Design, Engineering and Manufacturing actions aimed at optimization and balance. It is important to point out that as the functions remain the same to meet the requirements, the value graph does not change and all percentage movements must occur with each new cost proration generated by new solution proposals.

As recommended in the method presentation, the next step was to use DFA strategies in order to seek this optimization of solutions. To this end, the three classic questions were proposed for each component: i) Does the component or assembly being analyzed move in relation to the other parts in contact?; ii) Does the component or assembly being analyzed require a different material type in order to function?; and iii) Should the component or assembly being analyzed be dismantled or removed in any case of use? In this case, as they had parts assembled with all three "NO" answers, it was possible to suggest the unification of the parts, aiming to reduce costs of both raw material and labor for transformation, as shown in **Figure 9**, where the combination of solutions resulted in a cost reduction of R$ 4.35, that is, savings of 16%, with the perceived quality.

Comparing the graphs of the original solution (**Figure 8**) with those of the solution after the first iteration of the method (**Figure 9**), we can observe the approximation of the value and cost points by function. It is observed that the sum of the modules of the differences between value and cost has decreased, which increases the balance between the perception of value and the cost of industrialization.

With this first iteration, the results achieved have already shown a great improvement in the final cost. However, the proposed method suggests at least two iterations or iterations until the sum of the module of the differences between value and cost is less than 20% or when the DFA strategies no longer generate all negative answers for the three standard questions [13].

Thus, the three questions were applied again, at the end of which a new solution proposal never presented in previous discussions was reached. In this new design, a new improvement in cost results can be observed - a reduction of R$ 7.15, in relation to the original proposal, which represented a savings, now, of 27%, compared to the original solution.

However, more than the global cost reduction, there was a great approximation between the points of the value and cost graphs, function by function, as shown in **Figure 10**. Thus, through the comparisons of value x observed cost graphs in **Figure 10**, observing the three proposals tested, it is noted that the third - after the second round of DFA - was balanced due to the greater proximity of the points of each function. In addition to this clear better balance, it also resulted in a lower final cost.

Still in **Figure 10**, it can be seen that, as the functions do not change, their value graphs remain unchanged regardless of the applied solution. Thus, the cost graphs are adjusted by the points approach, function by function, where the sum of the differences of the value and cost percentage modules dropped from 59.2%, in the original version, to 34.0% after the first iteration and finally 18.1% after the second round of application of the DFA strategies, as foreseen in the proposed model. In addition to

Current solution - Steel body + "U" profile			
Functions	**Value (%)**	**Cost (%)**	**Difference**
A	1,3%	6,3%	5,0%
B	7,9%	9,9%	2,0%
C	46,1%	21,6%	24,5%
D	17,1%	17,1%	0,0%
E	7,9%	15,5%	7,6%
F	3,9%	11,9%	8,0%
G	15,8%	10,7%	5,1%
H	0,0%	7,0%	7,0%
		$\sum_{i=1}^{n} \lvert V_i - C_i \rvert$	**59,2%**

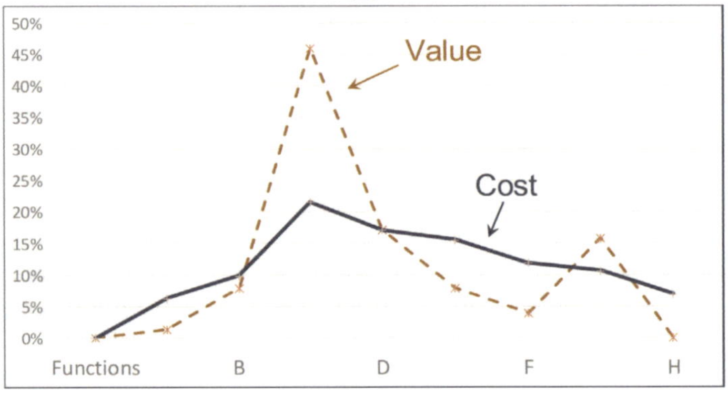

Figure 8.
Value x cost chart of original hinge assembly solution.

this balance achieved, another great gain came in the total cost of the solution, with a total drop of 27% at the end of the application of the proposal.

7. Considerations about the proposed DFB method

This chapter evaluated an integrated product development method that integrates the conceptual design and preliminary design phases. In order to seek a balance between the value perceived by internal and external customers for each of the functions of a set, with the cost of generated by the combination of solutions that meet these functions a new consumer good or part of it.

In this sense, the relevance of providing the multidisciplinary group of projects, which works in an integrated environment and simultaneously, with information on costs during the process of defining the solutions, to feed back the information of the conceptual design, seeking the balance value and cost. After all, the success or failure of the product in the field is largely determined by the initial development decisions. Thus, designers, engineers, represented in the areas of marketing, commercial, among

After iteration 1 - Body + spiked pin			
Functions	**Value (%)**	**Cost (%)**	**Difference**
A	1,3%	3,9%	2,6%
B	7,9%	8,1%	0,2%
C	46,1%	32,0%	14,1%
D	17,1%	15,8%	1,3%
E	7,9%	11,0%	3,1%
F	3,9%	8,4%	4,4%
G	15,8%	14,2%	1,6%
H	0,0%	6,7%	6,7%
		$\sum_{i=1}^{n} \lvert V_i - C_i \rvert$	**34,0%**

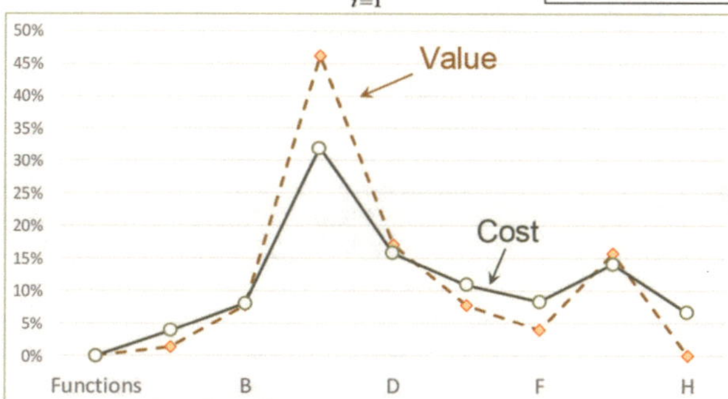

Direct Material + Labor (R$)	22,10

After iteration 1 - Body + spiked pin

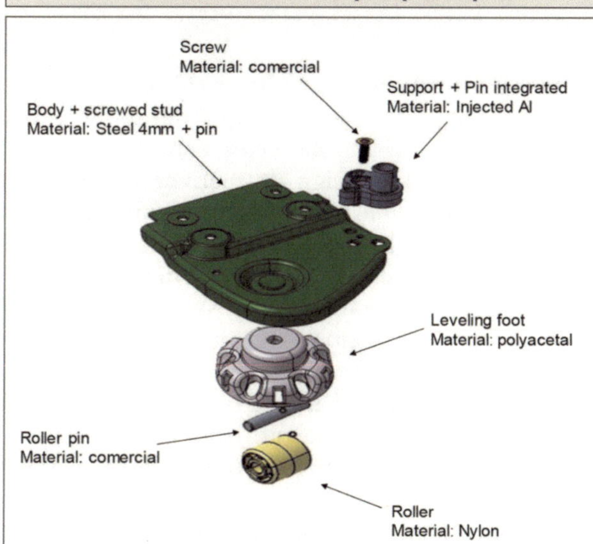

Direct Material + Labor (R$)	22,10

Figure 9.
Solution and costing after the first iteration. Value x cost graph of the solution after the first iteration.

Solução original - Corpo em aço + perfil "U"

Functions	Value (%)	Cost (%)	Difference
A	1,3%	6,3%	5,0%
B	7,9%	9,9%	2,0%
C	46,1%	21,6%	24,5%
D	17,1%	17,1%	0,0%
E	7,9%	15,5%	7,6%
F	3,9%	11,9%	8,0%
G	15,8%	10,7%	5,1%
H	0,0%	7,0%	7,0%
		$\sum_{i=1}^{n} \lvert V_i - C_i \rvert$	59,2%

Solução após 1° iteração: Corpo + pino cravado

Functions	Value (%)	Cost (%)	Difference
A	1,3%	3,9%	2,6%
B	7,9%	8,1%	0,2%
C	46,1%	32,0%	14,1%
D	17,1%	15,8%	1,3%
E	7,9%	11,0%	3,1%
F	3,9%	8,4%	4,4%
G	15,8%	14,2%	1,6%
H	0,0%	6,7%	6,7%
		$\sum_{i=1}^{n} \lvert V_i - C_i \rvert$	34,0%

Solução após 2° iteração: Corpo integrado em Al

Functions	Value (%)	Cost (%)	Difference
A	1,3%	2,9%	1,6%
B	7,9%	7,0%	0,9%
C	46,1%	40,3%	5,8%
D	17,1%	15,5%	1,6%
E	7,9%	8,0%	0,1%
F	3,9%	7,5%	3,5%
G	15,8%	15,0%	0,8%
H	0,0%	3,8%	3,8%
		$\sum_{i=1}^{n} \lvert V_i - C_i \rvert$	18,1%

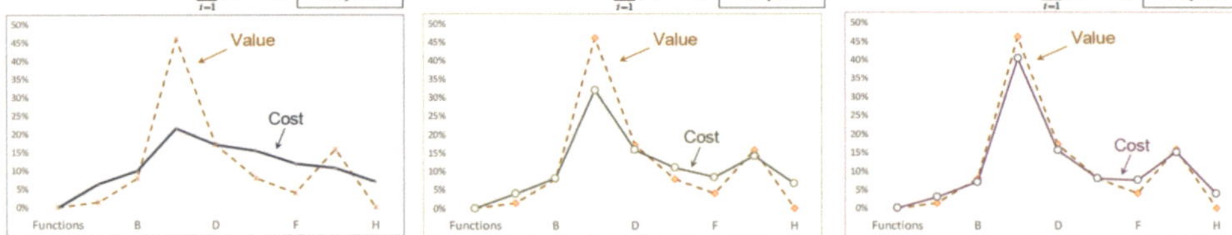

Material + Mão de obra (R$)	26,45	Material + Mão de obra (R$)	22,10	Material + Mão de obra (R$)	19,30

Solução original - Corpo em aço + perfil "U" | **Solução após 1° iteração: Corpo + pino cravado** | **Solução após 2° iteração: Corpo integrado em Al**

Material + Mão de obra (R$)	26,45	Material + Mão de obra (R$)	22,10	Material + Mão de obra (R$)	19,30

Figure 10.
Comparison of solutions, costs and comparatives value x cost of the original proposal and the proposals after the iterations.

others, must know the costs incurred in the phases subsequent to their decisions, and contribute to this search for balanced solutions, in order to meet their demands.

The model evaluated allowed the alternatives generated to solve the problems to be evaluated, stimulating the search for design alternatives, aimed at a balanced development between value and cost, while taking into account the requirements of the most diverse areas involved throughout the development product life cycle, including Marketing, Design, Product Engineering, Manufacturing, Logistics, Quality, Sales and Technical Assistance. Ensuring maximum service to your needs and committing the entire group to the product under development throughout its life cycle.

As a result, analyzing quantitatively, a total reduction of R$ 7.15 per product was observed. This means a direct savings of 27% in the final solution achieved when compared to the original combination. The verification of the costs in each round of the model application can be seen in **Figure 10**. In it is clear the importance and the gain of opportunities brought by the method. However, in addition to this quantitative analysis of the general costs, it was possible to observe a qualitative improvement in terms of the balance between the value and the cost of each function in the comparison of each cycle of application of the method. Still in **Figure 10**, it can be seen that the sum of the differences in the value and percentage of the cost, function by function, decreased with each iteration. In general, it is clear that the proposed method has significant qualitative advantages, as a better balance between value and cost increases the probability of project success. Likewise, the quantitative gains for this product are clear when considering an annual production of 200,000 units. Because, with this individual cost reduction of BRL 7.15 per product, annual savings of around BRL 1,430,000.00 are achieved just with this action in this specific subset.

As only a portion of the product was analyzed, further reductions in this family of consumer goods can be expected by expanding this analysis to other groups of parts and assemblies.

As a suggestion for possible future work, the model presented could be applied and validated in other product development models or even other types of productive organizations. Among these types, we could mention the automotive, cosmetics, food industries, among others. Other possible points of evaluation would be the analysis of the effect of applying this model on the other two supports of the economic-environmental-social triple bottom line, since the focus of this study was concentrated on the economic effects. After all, these changes in solutions also generated changes in materials and the ways in which people would interact with the product throughout its chain, from suppliers, through the assembly processes, to interactions with the final customer.

Author details

Paulo H.P. Setti* and Osiris Canciglieri Junior
PUCPR - PPGEPS, Curitiba, Brazil

*Address all correspondence to: settipaulo7@gmail.com

References

[1] Kim J, Park S, Kim HM. Optimal modular remanufactured product configuration and harvesting planning for end-of-life products. Journal of Mechanical Design. 2022;**144**(4): 1352-1361. DOI: 10.1115/1.4052389

[2] Unruh GU. Canciglieri Jr O. Identifying and Classifying Human-Centered Design Methods for Product Development. In: Ahram T, Karwowski W, Pickl S, Taiar R, editors. Human Systems Engineering and Design II. IHSED 2019. Advances in Intelligent Systems and Computing, vol 1026. Cham: Springer; 2020. DOI: 10.1007/978-3-030-27928-8_67

[3] Setti PHP, Canciglieri Junior O, Estorilio C. Integrated product development method based on value engineering and Design for Assembly concepts. Journal of Industrial Information Integration. 2021;**22**:21-45. DOI: 10.1016/j.jii.2020.100199

[4] Chan KY, Kwong CK, Wongthongtham P, Jiang H, Fung CKY, Abu-Salih B. Affective design using machine learning: A survey and its prospect of conjoining big data. International Journal of Computer Integrated Manufacturing. 2020;**33**(7): 645-669. DOI: 10.1080/0951192X.2018.1526412

[5] Soria Zurita NF, Stone RB, Onan Demirel H, Tumer IY. Identification of human–system interaction errors during early design stages using a functional basis framework. ASCE-ASME Journal of Risk and Uncert in Engineering System Part B Mechanical Engineering. 2020;**6**(1):1083-1098. DOI: 10.1115/1.4044787

[6] Sommer AF. Agile transformation at LEGO group. Research-Technology Management. 2019;**62**(5):20-29. DOI: 10.1080/08956308.2019.1638486

[7] Shen J, Erkoyuncu JA, Roy R, Wu B. A framework for cost evaluation in product service system configuration. International Journal of Production Research. 2017;**55**(20):6120-6144. DOI: 10.1080/00207543.2017.1325528

[8] Lu B, Zhang J, Xue D, Gu P. Systematic lifecycle Design for Sustainable Product Development. Concurrent Engineering. 2011; **19**(4):307-324. DOI: 10.1177/1063293X11424513

[9] Zheng C, An Y, Wang Z, Qin X, Eynard B, Bricogne M, et al. Knowledge-based engineering approach for defining robotic manufacturing system architectures. International Journal of Production Research. 2022:1-19. DOI: 10.1080/00207543.2022.2037025

[10] Marwa B, Mohamed A, Ikbal M, Vincent C, Mohamed H. KSim: An information system for knowledge management in digital factory. Concurrent Engineering. 2017;**25**(4): 303-315. DOI: 10.1177/1063293X17702689

[11] Zhang Z, Xu D, Ostrosi E, Yu L, Fan B. A systematic decision-making method for evaluating design alternatives of product service system based on variable precision rough set. Journal of Intelligent Manufacturing. 2017;**30**:1895-1909. DOI: 10.1007/s10845-017-1359-6

[12] Setti PHP, Canciglieri Junior O, Estorilio CCA. DFA concepts in a concurrent engineering environment: A white goods case. Concurrent Engineering. 2021;**29**:1-27. DOI: 10.1177/1063293X20985531

[13] Setti PHP, Canciglieri Junior O, Rudek M. Os benefícios da aplicação de iterações para avaliação de funções e para definição de soluções de design e engenharia no processo de desenvolvimento integrado de produtos. Design E Tecnologia. 2021;**11**(22):52-69. DOI: 10.23972/det2021iss22pp52-69

[14] Souder WE, Song XM. Contingent product design and marketing strategies influencing new product success and failure in U.S. and Japanese electronics firms. Journal of Product Innovation Management. 1997;**14**(1):21-34. DOI: 10.1111/1540-5885.1410021

[15] Winner RI, Pennell JP, Bertrand HE, Slusarzuk MG. The Role of Concurrent Engineering in Weapon Systems Acquisition. Alexandria: Institute of Defense Analyses Report R-338; 1988 1988

[16] Shouke C, Zhuobin W, Jie L. Comprehensive evaluation for construction performance in concurrent engineering environment. International Journal of Project Management. 2010; **28**(7):708-718. DOI: 10.1016/j.ijproman.2009.11.004

[17] Meng D, Yang S, Zhang Y, Zhu S. Structural reliability analysis and uncertainties-based collaborative design and optimization of turbine blades using surrogate model. Fatigue & Fracture of Engineering Materials & Structures. 2019;**42**(6):1219-1227. DOI: 10.1111/ffe.12906

[18] Chen B. Conceptual design synthesis based on series-parallel functional unit structure. Journal of Engineering Design. 2018;**29**(3):87-130. DOI: 10.1080/09544828.2018.1448057

[19] Pahl G, Beitz W, Feldhusen J, Grote K-H. Embodiment design. In: Wallace K, Blessing L, editors. Engineering design. 3rd ed, London: Springer; 2007. pp. 227-438. DOI: 10.1007/978-1-84628-319-2

[20] Chen SC, Huang JM, Yang CC, Lin WT, Chen RJ. Failure evaluation and the establishment of an improvement model for product data management introduced to enterprises. The International Journal of Advanced Manufacturing Technology. 2007;**35**(1–2):195-209. DOI: 10.1007/s00170-006-0705-1

[21] Prasad B. Product development process for IoT-ready products. Concurrent Engineering. 2020;**28**(2): 87-88. DOI: 10.1177/1063293X20932618

[22] Ullman DG. The Mechanical Design Process. 4th ed.; McGraw-Hill, ed. New York: Raghothaman Srinivasan; 2010 ISBN: 978-0072975741

[23] Unger Unruh G, Canciglieri Jr O. Human Needs: Analysis and Evaluation Approach for Product Development Context. In: Human Needs' Analysis and Evaluation Model for Product Development. Cham: Springer; 2023. DOI: 10.1007/978-3-031-12623-9_2

[24] Rauniar R, Rawski G, Morgan S, Mishra S. Knowledge integration in IPPD project: Role of shared project mission, mutual trust, and mutual influence. International Journal of Project Management. 2019;**37**(2):239-258. DOI: 10.1016/j.ijproman.2019.01.002

[25] Kim K, Lee K. Collaborative product design processes of industrial design and engineering design in consumer product companies. Design Studies. 2016;**46**: 226-260. DOI: 10.1016/j.destud.2016.06.003

[26] Shenas DG, Derakhshan S. Organizational approaches to the implementation of simultaneous

engineering. International Journal of Operations & Production Management. 1994;**14**(10):30-43. DOI: 10.1108/01443579410067234

[27] Lu P, Cai X, Wei Z, Song Y, Wu J. Quality management practices and inter-organizational project performance: Moderating effect of governance mechanisms. International Journal of Project Management. 2019;**37**(6):855-869. DOI: 10.1016/j.ijproman.2019.05.005

[28] Visser H, Thopil GA, Brent A. Life cycle cost profitability of biomass power plants in South Africa within the international context. Renewable Energy. 2019;**139**:9-21. DOI: 10.1016/j.renene.2019.02.080

[29] Yu E, Sangiorgi D. Exploring the transformative impacts of service design: The role of designer–client relationships in the service development process. Design Studies. 2018;**55**:79-111. DOI: 10.1016/j.destud.2017.09.001

[30] Drutchas J, Eppinger S. Guidance on application of agile in combined hardware and software development projects. Proceedings of the Design Society. 2022;**2**:151-160. DOI: 10.1017/pds.2022.16

[31] Project Management Institute. A Guide to the Project Management Body of Knowledge (PMBOK® Guide). 7th ed. Newtown Square, Pa: Project Management Institute; 2021 http://www.pmi.org. ISBN: 978-1-62825-184-5

[32] Harvey J, Aubry M. Project and processes: A convenient but simplistic dichotomy. International Journal of Operations & Production Management. 2018;**38**(6):1289-1311. DOI: 10.1108/IJOPM-01-2017-0010

[33] Ma H, Chu X, Xue D, Chen D. Identification of to-be-improved components for redesign of complex products and systems based on fuzzy QFD and FMEA. Journal of Intelligent Manufacturing. 2019;**30**(2):623-639. DOI: 10.1007/s10845-016-1269-z

[34] Setti PHP, Canciglieri Junior O. Método para Desenvolvimento Integrado de Produtos Baseado em EV e DFA. 1st ed. Vol. 1. Beau Bassin: International Book Market Service Ltd., member of OmniScriptum Publishing Group; 2018. p. 109 ISBN 978-620-2-19383-2

[35] Setti PHP. Modelo de Desenvolvimento Integrado de Produto Orientado ao Equilíbrio entre Valor e Custo de Funções [thesis]. Curitiba: Parana Catholic University; 2021

Chapter 4

An Approach to Improve the Product Design Process

Patrick Kenger

Abstract

The design work made by design engineers needs to be clearly defined and structured. The design process to follow should naturally lead to a systematic design work that reduces wasteful work, and reuses competence and knowledge from previous design work. In a time when products become more complex regarding performance, tolerances, material, and with more technological and digital things built into the products, this design process becomes even more important. The Way We Design (TWWD) presented in this chapter is one contribution to such a design process for the day-to-day design work. The work behind TWWD is based on both research and industrial practice over more than 15 years. The process is built on 11 checkpoints (A to K), which gives details on the design process and is customized for the specific company. Implementations of TWWD in companies are described in this chapter, where the results have shown a gain in lead-time, development cost, and improved design quality.

Keywords: product design process, lead-time, design cost, design quality, design complexity

1. Introduction

Product design and its process have a major role in the competitiveness of a product developing company. Long development lead times will extend the market introduction of new products. Many design hours will incur high development cost, which in the end will increase the market price of the product. Unclear design specification and testing will result in an uncompetitive product. Design defects during the design work will cause the need for rework, and in the worst case, the end customer will have a defective product. Defective products introduced on the market will harm the company's brand and market shares, as well as having a negative effect on the stock price. In the worst case, recalling the products or lawsuits will bring huge costs and negative media coverage for the company. See for example the summery made in table 5-3 p. 72 in [1]. Hence, a structured and effective product design process is essential for any product-developing company.

In a time when more technological and digital things shall be designed into products (memory cards, sensors, Bluetooth functions, I/O modules, radio units, displays, and a lot of software) it will be even more important that the baseline of how to design the products are defined and fully understood. Companies that we have been working with in research projects show that one does not have enough routines for the design work.

For example, they lack in their work for how to specify the design work, manage standards and directives, dimension parts, how to make CAD parts and assemblies, review drawings, make tests, or manage their documentation. These companies will have an even greater challenge to stay competitive now when products become digitalized, and even more demands are introduced into the design work.

Hence, the requirements of a structured and effective product design process will become even more essential due to digitalization than it has been so far. The whole product design process as well as the products will become more complex. Which put very new demands on the way the company develops their products.

2. Product design processes in general

The research regarding product design processes itself has been going on at least since the early 1960s when the work by [1] is commonly cited. Even though [2] mentions that the earliest work reported is from 1948. During the 1960s, the design process was poorly developed and was far behind the manufacturing process development. Norden [1] stated that the design work was a matter of art and craftsmanship, rather than science and system. The design work was at this time typically known for its creativity and uncertainty. The first conference on design methods was held in London in September 1962 [2], which is known as the starting point of assembling researchers within the area.

Since the end of the 1980s, product development processes and methods have been available to the industry. Some of the known and commonly cited literature that describes theories and processes in product design can be found in [3–7]. These authors have made an important impact on the research community and the education of engineers at universities worldwide. Furthermore, in 2000, [8] launched their book product design and development which since has been renewed each 4th year. The book summarizes the work of other researchers and contributes to the concept development process.

At the beginning of the 1990s, modular design and product platform development emerged. However, companies have worked with modular product development since the 1930s. Scania, for example, initiated its modular development of truck components in 1939 [9]. IBM was a forerunner with their modular computer in 1964 when the System/360 was launched [10]. The underlying ideas regarding standardization and modularity started already in 1914 in the automotive industry [11]. One may even see Taylor's work in the early 1900s as a predecessor to both development processes and modularisation [12]. His ideas on standardized and repetitive work have analogies with today's reasons why companies develop modular products, for example, common units and carry-over modules, see [13]. Also, the work by [14] can be seen as one of the initiators of modular thinking. Modularisation has been part of many companies' way of designing products for the last 20 years. Many of them are described in Refs. [11, 13, 15, 16].

A comprehensive review of the available literature regarding product development processes is made by [17]. The one really interested in the wide range of processes; their work will guide further to a numerous of interesting literature. In [17], they classify different design and development process models into micro-, meso- and macro-level. Based on these levels, they also classify the design and development processes into procedural, analytical, abstract, and mathematical/computational. Their findings in [17], which one might agree upon, is that most situations for a product developing company require more than one category of model. In addition, the findings from reading their work, which they also conclude, is that the process should be built for its purpose. That is, the product design process should be made considering the company's needs.

During the last two decades, the lean product development (LPD) methodologies have been developed. Based on the lean manufacturing methods, as described in [18–21], lean product development aims at identifying wasteful design work and doing things as streamlined as possible. This has put a well-deserved focus on the design process. Some of the LPD work is based on, or inspired by [19, 21, 22]. As with lean manufacturing, LPD should identify and remove wasteful processes or activities, and follow certain principles. For example, [22] describes the lean way of Toyota which is built on three main areas; people, process and tools, and technology. Within these three areas, there are 13 principles to follow and apply to become leaner. Toyota is also the base for the book by [23] where the knowledge-based design are developed and described; which is another way of becoming a lean product development organization.

Different approaches have also been made to improve the design processes regarding specific areas. Improving the testing process and integrating it into the product design process [24] highlights some important factors to consider. For example, the design complexity, architecture, and novelty of the product, as well as the timing of the test itself. They conclude that it is important to depict testing more explicitly in the product design process.

To further improve the integration of industrial design and engineering design [25] looks into the potential ways of collaboration. In their work, foremost based on literature reviews and interviews, they propose a model for collaboration. They also mean that more concrete process models are needed for the specific company. How companies are working to improve the environmental impact of new product development was studied by [26]. Specifically, 10 companies were studied how well they have adopted ERM (environmentally responsible manufacturing) in product development. One of the outcomes of their work is that to design products environmentally friendly, the environmental aspects need to be considered early in the product design process.

3. Improvements in the product design process

During the work to develop The Way We Design (TWWD) and the implementations at companies, it is evident that despite the research and available literature on product development. Many companies still struggling to get a competitive product design process. The author's own experience is that this depends on organic growth of company culture, just as described in [22], this culture is difficult but essential to change. It is built in into the day-to-day design work. New employees just follow old habits, instead of bringing new knowledge and ideas into the company. Still, most of the design engineers, design managers, and management team are well educated and should know better ways of designing things.

A lot of research has been produced within the area where [17] mentioned a lot of the literature. Still, models of different levels and abstractions are not made explicitly for industrial product design purposes. Within the research community, the challenges with product design and development processes are well recognized. Where one of the well-known journals within the area made a special issue on the topic. In that special issue [27] says that there is a need to continue to research the area of design and development processes since better processes will bring better products to the market and thereby increase profit to the company.

Furthermore, research indicates that the design work has a huge influence on the internal manufacturing and assembly processes, as well as the final product cost. Work by [28] concludes that there is little evidence of how much this influence actually is. However, the cost of the manufacturing and assembly process is partly a direct result of the design work. Here [29] indicates that the design work determines 47% of the manufacturing cost, at least for a case at a coffee machine manufacturer. In this context, it is even more important that the design work is made according to a process that includes the demands from manufacturing and assembly, as well as other internal processes at the company.

The methods and processes available in the literature are on a high and abstract level, or they just cover a small piece of the complete design work and are therefore not suitable for a direct implementation at a company. In addition, available methods and processes assume that a new product shall be designed. While much of the design work at companies is just small changes in parts, drawings, or in part number structures. For companies, the time and work to change their way of doing things seems too long and undefined. As a result, the company and design engineers keep on doing things in the same (poor) way as before.

Also, it is not uncommon that the software design, electrical design, and mechanical design are working separately from each other, in their own way of designing. This occurrence can be seen in practice at three companies. Each of them even had separated group managers, which in turn had their own view on product design. When the products are embedded with all kinds of new technology, and the company goes from selling a product to offering a service, the product design process needs to integrate all key engineering functions to design a competitive (and complex) product.

Companies that suffer today from a poor product design process will have an even greater challenge to stay competitive in the future. Hence, when products become more digital, and more demands are introduced into the design work, the design process will be more challenging. As a result, the requirements of a structured and effective product design process will become even more essential than it has been so far.

Given the literature and examples discussed above, it is justified to develop a design process that is detailed enough for design engineers to understand and adapt to, and which describes what and how they should do things. This design process shall guide the design engineers in their day-to-day work and include all essential parts such as legalization and standards and be possible to monitor and manage during any design project. This design process should also make the design engineers work as an efficient whole, and not sub-optimize based on individuals. TWWD is one such design process that will contribute to improving the product design process.

4. The way we design process

The objective of the TWWD process, see **Figure 1**, is to provide a structured and efficient product design process ready to be implemented in an industrial company.

DOI: http://dx.doi.org/10.5772/intechopen.107427

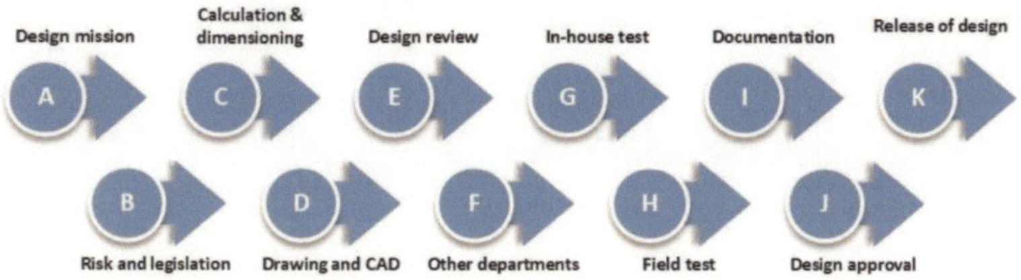

Figure 1.
The Way We Design (TWWD) process with K number och checkpoints.

The objective has also been to make it detailed enough so that each designers know what to do when following the process; but still a generalized product design process that can be implemented to a company. Most important; the objective with TWWD is a design process that will predict the outcome from the designer's work and improves the R&D department's possibility to meet cost-, quality- and lead-time targets.

TWWD process needs to be owned and managed by the company, not by the individuals in the staff. Consequently, the knowledge of how to design the products shall be the core competence of the company. Therefore, as a part of developing the TWWD, continuous improvement and the management of the process have been considered.

The checkpoints in the TWWD process (A to K), see **Figure 1**, are essential to a company, and can seldom be ignored. Hence, the TWWD process is built on fundamentals applicable to any company.

5. The fundamentals of TWWD

TWWD process is built on four fundamentals: (1) checkpoints and the design manual, (2) documents and TWWD files, (3) the TWWD checklist, and (4) training material and TWWD management. Altogether, this builds up the TWWD process in a structured and efficient way.

5.1 Checkpoints

The first fundamental in TWWD is the checkpoints. The checkpoints range from checkpoint A (when the design work starts) to K (when the design work is completed), see **Figure 1**. Each checkpoint contains a question that needs to be considered by the designer, and answered with "**Yes**", "**No**", or "**Not applicable**" (**N/A**). To answer the question with **Yes** means that the checkpoint was necessary to complete, and the work was performed with a positive result and at an appropriate quality.

To answer the question with **No** means that the checkpoint was necessary to complete, but the work was not performed at all, or did not comply with the intended quality. In this case (when answering **No**), the responsible person needs to approve the deviating work, or the checkpoint is reworked and completed. To answer the question with **N/A** means that the checkpoint is not necessary to complete and does not apply to the design work in regard.

Each of the checkpoints A to K contains a set of sub-checkpoints, e.g. E1, E2, and E3, see examples in **Figure 2**. However, the day-to-day design work in the design of parts and products are unlikely to include all the checkpoints in the process. A major

Figure 2.
Checkpoint E and its sub-checkpoints.

part of the design work may be small design changes and updates. Therefore, the checkpoints are many times N/A and not necessary to complete.

The design engineers do however work according to TWWD without continuously reading the design manual describing TWWD. Nor do they necessarily move forward in a stepwise manner, checkpoint by checkpoint. As TWWD becomes implemented and familiar to the designer, the work is carried out in a smooth and lead-time efficient way. Then TWWD acts as the red-line and the base for how the designer shall design parts and products.

The checkpoints are not of equal magnitude and require a different amount of work to complete. In addition, the different checkpoints may be completed by different persons due to the nature of the checkpoint. The sub-checkpoints are arranged so that each sub-checkpoint (e.g. B1, B2, B3, etc.) can be completed by the same person. That is, risk assessments in B can be made by one person(s), while calculations in C are made by another person(s). Furthermore, to complete the design reviews in E might not be the same person(s) as performs the field tests in H. Hence, each checkpoint is characterized by a certain engineering skill.

Altogether, the checkpoints in TWWD are arranged based on the systematic nature of the design work, and the involved personnel required to complete the checkpoints and sub-checkpoints.

5.2 Checkpoint example—checkpoint E4: electrical design review

The complete TWWD process contains 11 checkpoints and 120 sub-checkpoints. These are not possible to describe in this chapter. However, for the reader to be able to follow and understand the rest of this book chapter, sub-checkpoint E4 is here written in its complete just as is written in the TWWD process.

5.3 Sub-checkpoint E4

An electrical design review is made to assure that the electrical design work and the designed parts and components are made correct. Always make an internal design review for work that is considered costly or comprehensive. Personnel who is part of a general design review meeting is the designer, an experienced R&D college, and personnel at other departments relevant for the review. The designer in charge of the design work assembles the personnel to perform the design review.

Checkpoint question: Is an electrical design review completed?
To answer yes to this checkpoint:

- The designer shall send out drawings and/or material at least 2 working days ahead of the planned review meeting.
- The designer shall assemble suitable personnel for a general design review meeting.
- The designer shall present the designed part and its performance using drawings or 3D models, electrical schematics, or prototypes.
- When applicable, the designer should present several possible solutions (concepts) and together with the attending personnel select the best possible solution.
- The designer shall secure that the design fulfills the design specifications and performance (set in checkpoint A).
- The designer shall, based on the design review, make suitable adjustments to the part or design.
- Is the material/standard component available for the factories in China (or other preferable countries)?
- The designer shall document the general design review meeting using the TWWD review template in the TWWD files.
- The designer shall file the document according to documentation requirements.

5.4 Checkpoints and a good enough design

To overwork, the design of the part or product is not value-adding for the customer or the company. If the design is too poor, this will harm both the customer and the company. The design of the part and product shall be good enough, see **Figure 3**.

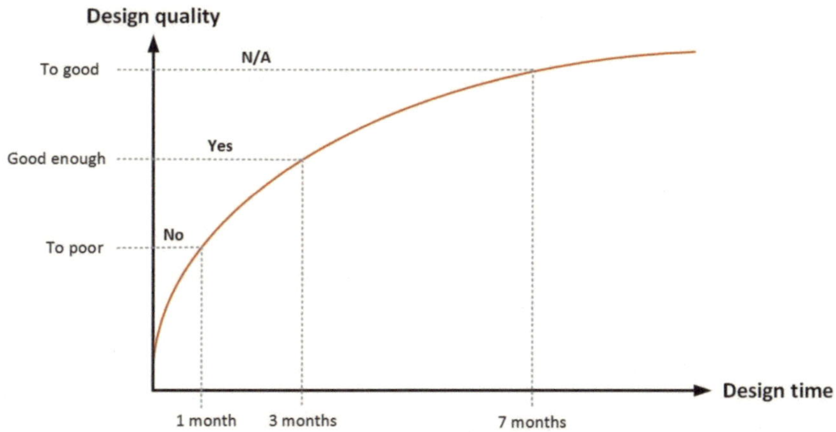

Figure 3.
A good enough design and the answers to the checkpoints in TWWD.

Good enough means that the product fulfills all demands from legalizations, standards, directives, customers, and our company-specific requirements, not more and not less.

The design quality graph in **Figure 3** illustrates the idea that it is not a linear relationship between design quality and design time. At some point, the quality improvements are not worth the effort in hours. The quality improvements are not possible to charge the customer, and the customer is not asking for this additional quality. For example, twice the needed life length of any given product. As a result, the design work needs to target the right quality for the right cost, which the customer is asking for and is willing to pay for.

The answers to each checkpoint in TWWD (N/A, Yes, and No), will support the design work to be good enough. If a checkpoint is answered No, the design quality will be too poor (since the checkpoint should be completed). If a checkpoint is answered Yes, the design quality is good enough (since it is necessary to complete the checkpoint). If the checkpoint is answered N/A, we have avoided that the design quality would be too good (since the completion of a N/A checkpoint would create an overworked design).

5.5 The design manual

In addition, as the first fundamental in TWWD, the design manual describes all checkpoints and how to complete them. The actual manual is a company-specific document that describes the checkpoints and how to complete them. The "company-specific" refers to the unique description of each checkpoint and the unique way to complete each checkpoint for the specific company. Hence, TWWD needs, to some extent, to be customized for the specific company and their "The Way We Design". The design manual is a digital document stored at a specific place according to the documentation requirements; and/or a physical booklet was given to each designer in order to simplify its compliance during the day-to-day design work.

The results by following the design manual shall be documented (checked) in the TWWD checklist for each completed design work. The results shall also be documented according to any specifications at each checkpoint, and according to the documentation requirements at the company. It is the designer's, or the involved personnel's, responsibility to answer and complete the checkpoint based on the best possible judgment and after performing as good as possible for the specific checkpoint. It is understood that all designers perform as good as possible and with the best possible judgment. After all, they have been employed at the company based on their engineering skills and personality. Hence, they are entrusted to have a good judgment and to follow the TWWD process and the design manual.

The designer shall not overwork each checkpoint or make it more complicated than it is. On the other hand, do not try to complete each checkpoint as fast as possible and jeopardize quality, performance, and total design lead-time (since the design may need rework if things go too fast which results in a design error). If something in TWWD is not clear, the designer needs to discuss this with the colleague or manager. If something in TWWD shall be overruled, the designer needs to discuss this with the manager for decision and action; or with the personnel that has the authority to decide on the specific checkpoint.

5.6 Documents and TWWD files

The second fundamental in TWWD is the documents and the TWWD files. There are many reasons why documents, templates, and the management of the documents shall be a defined part of a design process. One reason is the legal demands on documentation that are described in directives such as the directive of machinery [30]. Other reasons are the internal knowledge and competence that are documented in reports and guidelines. TWWD includes the most common documents and templates that are required to store, maintain, and reuse knowledge gained during the design of parts and products.

In checkpoint I, TWWD manage documents, as well as documentation requirements for each document. The TWWD files are the name of the folder system for all TWWD documents. The TWWD files contain the support documents needed for the TWWD process at specific checkpoints. Some companies store documents in many different locations, including local at individual designers' hard drive, as well as in SmarTeam or other PDM systems. During the last few years, SharePoint has become a commonly used base for document handling in the industry. However, the most important is to agree on where and how all documents shall be approved and stored.

5.7 TWWD checklist

The third fundamental is the TWWD checklist. At checkpoint A, the design work is started, and at checkpoint A2 a new TWWD checklist is established. The checklist is digital or in a paper format (depending on the way the company prefers) and follows the design work, checkpoint by checkpoint. When the design work is completed at checkpoint K, the checklist is stored according to the documentation requirements. For a project manager, the checklist also acts as a project planner and status report. The project manager can manage the project by reviewing each design engineer's progress in the TWWD checklist. As well as each designer in a project can tell where in the design process he or she is, and what remains before the release of the design in checkpoint K.

5.8 Training and management

The fourth fundamental in TWWD is the training material and management of the process. To train the designers and personnel at other departments in the company, a training material should be developed. The design manual together with a set of powerpoint sessions serves well as the training material. This training material describes and explains the different parts of TWWD. This is important, as a part of continuous competence development inside the company, as well as to train new employees.

Finally, the R&D manager (or the assigned person) shall be responsible for the governance and continuous development of the TWWD process. To this person, other personnel can submit suggestions for improvements to the company's TWWD. The TWWD manager will manage the process and is responsible for its maintenance, major updates, and that the process is followed including arranging training sessions. The R&D manager can delegate the responsibility and management of the process when TWWD is becoming the regular way and the designers are fully adapted to the process. Then the TWWD management can be delegated to another suitable person for the continuous support and maintenance of the process.

It is important that the designers contribute to developing TWWD for the company by bringing forward things that work poorly, and things that they believe are missing. Specifically, the designers should not hesitate to propose new and better things that can update TWWD. The company should continuously improve TWWD; this will give better prerequisites for the company to be even more competitive. Specifically, after completing checkpoint K, the designer is asked to document any improvements for TWWD and submit this information to the TWWD manager at the company.

6. Concurrent set up of TWWD

One important thing to accomplish short lead times in product design is to work and complete the activities concurrently. Here, the term concurrent is used but also the term "integrated" or "simultaneously" are widely used terms. The overall meaning is that things are done at the same time, in parallel, instead of in a serial pattern. The concurrent way to work has been known for decades to be a good way of planning and completing different activities. The concurrent work to design a product and its production system is thoroughly described in [31]. Their pioneering way to describe the ideal integration of design and production is aimed at better handling the increased complexity of manufacturing, competitiveness, new technologies, and the actual product and its design. Another pioneer within the concurrent set-up of work tasks is the work by Andreasen in [32]. Specifically, he describes the benefits and ways to integrate the areas of market, product, and production. Andreasen argues that product development cannot be made in the best possible way without integrating different areas of specialization. It is easy to agree on this, for example, that different ongoing product development projects need to integrate with each other. Alternatively, that different people with different responsibilities within a company need to agree on the best possible solutions. Hence, just focusing on increased assembly efficiency is not the only important thing when designing the product. Even though the work by [31, 32] was written more than 30 years ago, the concurrent way of working is more valid than ever. Hence, a company needs to be as efficient as possible in whatever they are doing. Today, concurrent design is widely used[1] by both researchers and practitioners to increase the efficiency of the design process and activities.

To optimize the lead time in TWWD, the checkpoints should be completed concurrently, when possible. The illustration of the TWWD process in **Figure 4** shows that the design work is performed by starting at checkpoint A and then finalized in a consecutive way onwards through the checkpoints. Checkpoint A is completed before starting checkpoint B, and so on. Note the importance, the process is designed and illustrated in such a way that each checkpoint is the last point of completing the design

[1] A search at google scholar on "integrated product development" will reveal 1,6 million hits. Since 2018, more than 50 000 text on the subject has been produced. Applied at a wide range of business and research areas.

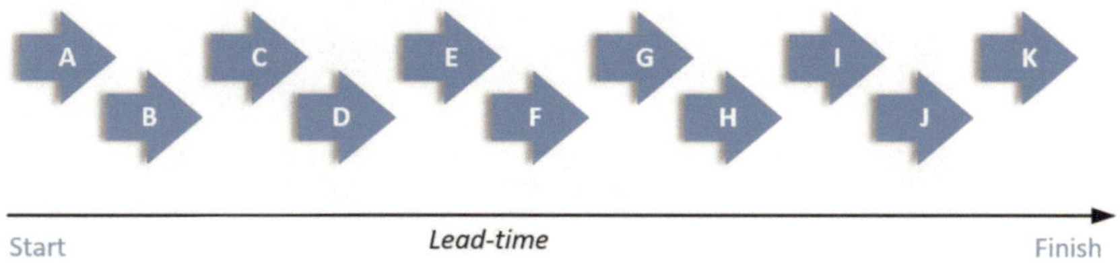

Figure 4.
Illustration of the TWWD process. The design work is performed by starting at checkpoint A and then finalized in a consecutive way onwards through the checkpoints. The checkpoints denote the last point of completing the work.

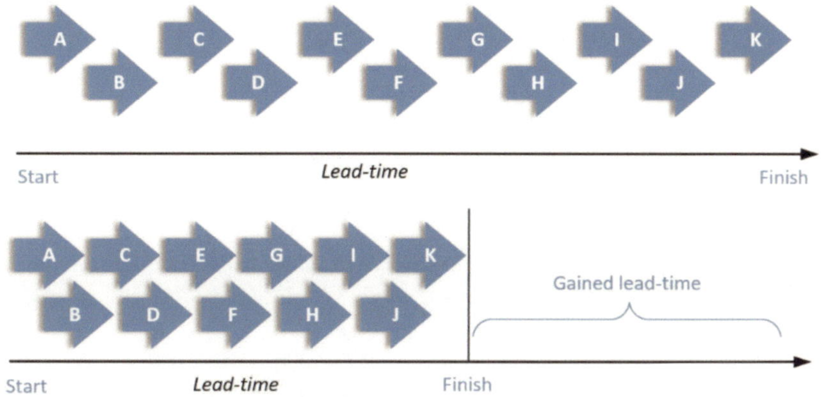

Figure 5.
The planning and the completion of TWWD will gain in lead time when the checkpoints are concurrently completed.

work. Hence, the checkpoints, and sub-checkpoints in the process denote the last point in time when the checkpoint should be completed.

Concerning the illustration in **Figure 4**, to optimize the TWWD process and design work, one shall plan the work so that less time (or no time) is spent on idling between the checkpoints. Practically, that means the designers and project manager carefully plan the design work so that any handover to other designers, as well as the startup of the next checkpoint, are made at the very best possible time. In this way, as illustrated in **Figure 5**, lead time will be gained.

In the upper part of **Figure 5**, the original illustration of the process shows that there is a gap (idle time) between the checkpoints. In the lower part, the TWWD process is compressed by having a well-planned process. In the lower part of the figure, each checkpoint is completed concurrently and handed over in a time-efficient way. At the end, a well-planned process will gain in lead time, compared to a process that is poorly planned.

For a specific checkpoint, for example, checkpoint B, not all sub-checkpoints may be needed for the specific design work. Hence, some sub-checkpoints can be removed based on the nature of the development project and the design work required. In addition, each of the sub-checkpoints may be integrated with each other to decrease the lead-time in the design work. Removing the sub-checkpoints that, for sure, is not part of the design work will then reduce the design lead-time.

As a result, of a well-planned TWWD process, one may identify, already from the start, which checkpoints can be removed from the process. In **Figure 6**,

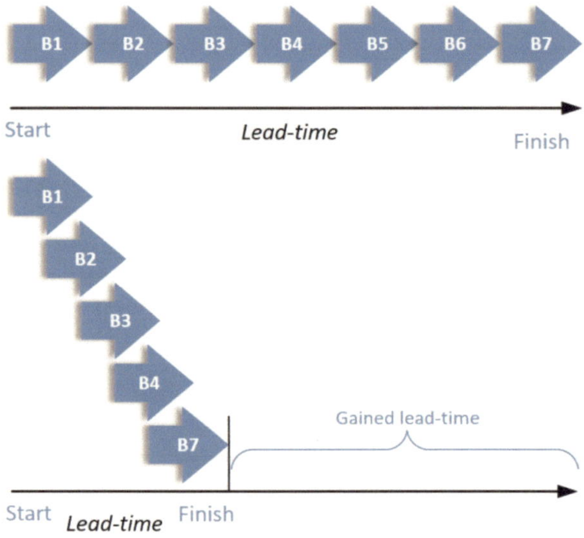

Figure 6.
Removing sub-checkpoints that will not be part of the design work, as well as integrated each sub-checkpoint will reduce the lead-time for each checkpoint.

sub-checkpoints B5 and B6 have been removed from the design work. In the same figure, the sub-checkpoints have been integrated with each other. Meaning that each sub-checkpoint is started before the previous sub-checkpoints are finished. Specifically, this is possible to accomplish when different design engineers are involved in completing the sub-checkpoints. In the end, further gains are made in shortening the lead-time in the design process.

6.1 Structuring the concurrent work in TWWD

Planning, compressing the design work, and integrating the work activities will bring gains in lead-time. To further optimize the design work in TWWD, one should identify dependencies and relations between the different checkpoints. As well as structuring the checkpoints in the most suitable order for completion. One way to do this is to use the design structure matrix. The design structure matrix (DSM), as described in [33–35], is a good way to structure the design work. The DSM was originally described in [36]. The article shows how DSM be applied to organize the design work and how to accomplish an effective development plan. Specifically, to identify and analyze the dependencies in a development plan. By doing this, one can identify possibilities to restructure the design work and thereby reduce the lead-time as much as possible.

At the very start of TWWD, it is the project manager, or any other that will manage the design work in TWWD, that will plan the order for the completion of the checkpoints. Depending on the company, product, and design work, this plan will be different for each of the design projects. In the DSM for TWWD in **Figure 7**, each checkpoint is structured in consecutive order. Just as illustrated in **Figure 4**, the next checkpoint is started when the previous is completed. In the matrix, the rows are the checkpoint to complete. The columns are the checkpoint that is required to be completed before the next checkpoint in regard can be started. The "X" in the DSM marks

Checkpoint required as input (columns); Checkpoint to complete (rows)

	A	B	C	D	E	F	G	H	I	J	K
A	X										
B	X	X									
C	X	X	X								
D	X	X	X	X							
E	X	X	X	X	X						
F	X	X	X	X	X	X					
G	X	X	X	X	X	X	X				
H	X	X	X	X	X	X	X	X			
I	X	X	X	X	X	X	X	X	X		
J	X	X	X	X	X	X	X	X	X	X	
K	X	X	X	X	X	X	X	X	X	X	X

Figure 7.
Design structure matrix for TWWD. Without any restructuring of each checkpoint.

this relation and indicates which column (checkpoint) need to be finished before the next row (checkpoint) can be started.

As shown in **Figure 7**, the baseline for TWWD indicates that B cannot be started before A is completed; C cannot be started before A and B are completed, and so on. However, by planning for each product design project and restructuring the work, one may further cut the design lead-time and complete each checkpoint as just at the right time. In **Figure 10**, the DSM for TWWD is illustrated. The DSM is arbitrary but still a typical situation when checkpoints are restructured based on the actual opportunities in the product design work. In the DSM, no field tests are required so checkpoint H is removed.

In addition, the documentation required to complete checkpoint I does not need input from checkpoints F, G, J, or K. As a result, checkpoint I is restructured to be completed after checkpoint E.

The DSM in **Figure 8** shows another example where there are several spots where there are no dependencies. For example, any work is not required in checkpoint K until J is completed. For the project manager, to develop a DSM for the project ahead, it is a support to staffing the design engineers to the right work activity at the right

Checkpoint required as input (columns); Checkpoint to complete (rows)

	A	B	C	D	E	I	F	G	J	K
A	X									
B	X	X								
C	X	X	X							
D	X	X	X	X						
E	X	X	X	X	X					
I			X	X	X	X				
F	X	X	X	X	X		X			
G	X				X		X	X		
J					X	X	X	X	X	
K									X	X

Figure 8.
An example of a DSM where some independencies have been identified. Checkpoint H, field tests, has been identified as not required. Therefore, checkpoint H is not part of the DSM. As well as checkpoint I has been restructured and can be completed after checkpoint E.

Checkpoint required as input

Checkpoint to complete	B1	B2	B3	B4	B7
B1	X				
B2		X			
B3			X		
B4				X	
B7	X	X	X	X	X

Figure 9.
The design structure matrix applied at checkpoint B. In this example, checkpoints B5 and B6 are not required for the design work in regard. In addition, B1 to B4 are independent of each other.

time. Specifically, this is useful when the design engineers are enrolled in several projects at the same time.

The DSM can be applied at each checkpoint as well. To exemplify this, in **Figure 9** a DSM has been made for the completion of checkpoint B. For the specific product design work in checkpoint B, checkpoints B1, B2, B3, and B4 can be made simultaneously and independent of each other. In its DSM, this is indicated by no X-marks in the intersection of rows and columns between B1 to B4. However, B7 cannot be completed before all other sub-checkpoints are completed. The DSM for checkpoint B shows that there is an opportunity to complete B at a relatively short lead-time. This is given four design engineers available, that can work on each of the sub-checkpoints simultaneously.

7. The starting point of TWWD

When is then the starting point of TWWD then? As one may have noted, TWWD starts with checkpoint A—the design mission, see **Figure 1**. TWWD is a product design process, meaning that it does not by far cover all parts of the product development process. TWWD starts when the product design mission is set. As illustrated in **Figure 10**, comprehensive work is assumed to be completed in the product development process before the start of checkpoint A.

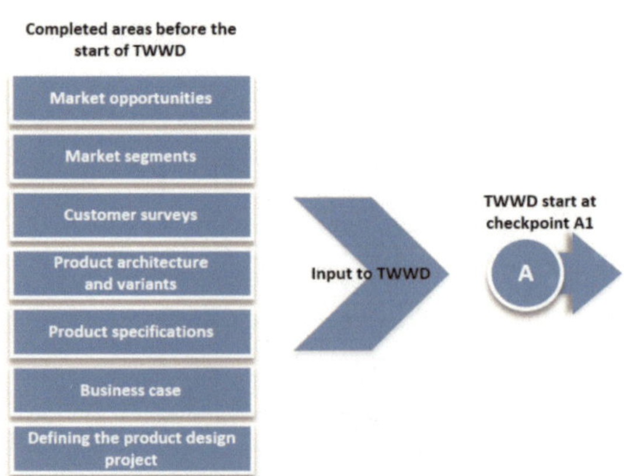

Figure 10.
The areas and work assumed to be completed before TWWD; and are given as input to the start of TWWD in checkpoint A.

For example, the opportunities the company have identified by launching a new product; or any segmentation of market and users have been completed. Also, customer surveys and needs have been identified; as well as any specific architecture and product variants have been set. The specifications for the product and its variants have been set (at least in overall terms). The business case has been identified including the target product cost. In addition, prior to the start of TWWD, the product design team has been identified and assigned to the forthcoming design work.

When the product design project is defined, then the design specification arrives at the design department who then will complete the actual product design. At this point, the TWWD will come into play and set off at checkpoint A1.

Worth mentioning, for some companies much of the product design work are for small upgrades and redesigns based on identified defects and errors, or planned upgrades. For these situations, TWWD can be applied at once and does not have to await all areas in the above figure to be completed again.

8. Improving the product design process in a vehicle electronics company

Several implementations of TWWD have been made in Swedish and Asian companies. In order to describe the implementation, work as well as potential gains, this section describes the actual implementation in a vehicle electronics company.

The work to implement TWWD was made from 2018 to 2021. Although the actual time for the implementation was approximately 18 months. The implementation was really challenging since there was a rather turbulent period for the company, as well as globally due to the Covid-19 pandemic.

The initiative for the TWWD implementation was a general need for being even better than today regarding product design and its lead-time and quality. The company has grown relatively much for some time, more design engineers are entering the company, as well as more and more product variants are offered to the market. The company's R&D is built on electrical, software and mechanical design. The implementation team was initially seven persons and the R&D manager. Of which, there were two design engineers from each of the three engineering areas. An initial plan was made with the aim of having the implementation done in 49 weeks. The set-up for the teamwork meetings was that we met every other week where we discussed each of the sub-checkpoints one by one.

One of the first things that were done was to sketch an overview and description of the way the company manages documentation and the routines for who will review and approve the documents. During the discussion regarding the documentation, it was clear that this would be an important thing to improve the design work at the company.

After some months of work, the company purchased another company. This meant a change in the implementation team composition. New design engineers joined the team and others left the team. Hence, there was a need to get the new team members up to date with the work so far. This was done by separate meetings with the new team members where the author tough the TWWD and explained the work so far. Still, we were working as a cross-functional team with design engineers from all three areas. At these meetings, the R&D manager participated, as well as key persons from software-, electrical- and mechanical engineering.

Specifically, during the implementation work, it was clear that there were three separate minor development ways at the company. One way for electrical design, another for software design, and a third way for the mechanical design. In order to fully understand these three ways, we had separate meetings with each of the three. At these meetings, it was clear that within all three areas, there were also individual ways of doing things. One of the biggest challenges was then to get all together and work similarly toward an efficient whole.

Another challenge at the company was that the complexity of the product made it rather difficult to manufacture. In addition, mechanics, electronics, and software had to work perfectly as an entire. Due to the use and the contents of the products, there are many regulations that the products should comply with. This further stressed the importance of being even better at managing documents, and the requirements for those.

Due to the continuous growth of the company as well as the newly acquired company, the implementation work was set on held. Approximately 6 months later the implantation work was back on track. Meanwhile, another company had been acquired, as well as even more growth in Asia. Yet again, new team members joined the TWWD implementation, as well as other left. There was also a new R&D manager in place. The new R&D manager together with a new implementation team improve the work as well as the stress to get TWWD in place. Due to the rather turbulent period, the growth, and staff turnover, it was even more important to get a common way of designing products.

Well up and running again, a new and good way of working was developed. Before each meeting, the design engineers and the R&D manager really studied the forthcoming sub-checkpoint. As well as figuring out how the checkpoint relates to other know and good routines that should be kept and integrated into TWWD. In addition, the top management was encouraging the implementation to push through and to be finalized. With the rather new way of working, each implementation meeting was fast and fruitful and we could focus only on the actual tuning the sub-checkpoints in regard.

Just as we were about to reach checkpoint I in the implementation work, the Covid-19 pandemic hit Sweden hard. As for most of the companies around the world, orders went down, and personnel were sick as well as working at home. The company put the implementation work again on hold. We did a few attempts to start up again, but we were put on hold. New ways of having meetings through Microsoft Teams or other video conferencing software made it possible for us to get back on track again.

In November 2020, we started up the implementation work through video conferencing, now, with a smaller group of attending design engineers. We rescheduled the implementation several times and we had a few more breaks during the final checkpoints. However, despite all challenges, we finally completed the implementation according to the implementation plan in **Figure 11**.

8.1 The investment cost

During the approximately 18 months of actual implementation work, there were an average of six design engineers, including the R&D manager being part of the work. On an average, the engineers worked two hours per week. This gives a total of 864 hours of work. At an hourly rate of 700 SEK, the total investment cost is 604 800 SEK.

8.2 Payback calculation

Among the benefits was that the early preparation stages before starting a project do have more substance, better confidence, and thus decreased risks of late and costly

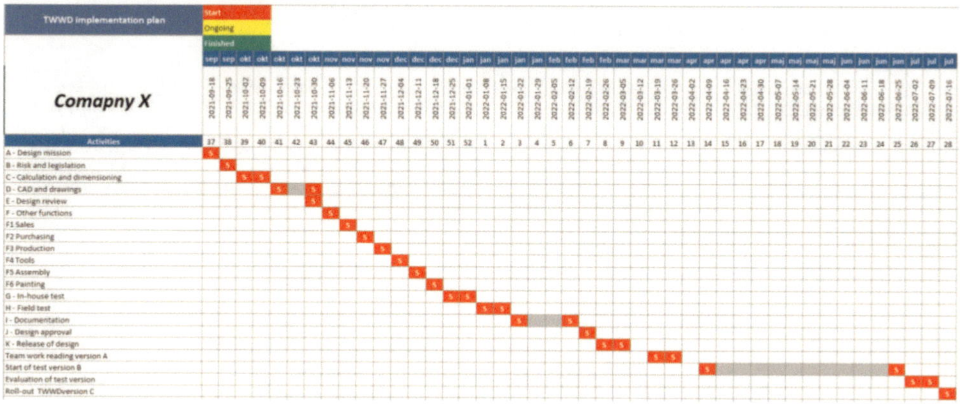

Figure 11.
The final implementation plan at company X (Vehicle Electronics Company).

change requests. In short, checkpoint A helps the company to be as ready as possible before starting the design work. Second, the obvious benefit of having a clear and structured way of working are to be able to see where the designers are working at, which checkpoint they are. Everyone knows what is expected from them, and engineers can work in the same way. Third, the company's documentation is more structured and harmonized which makes them work according to standards and legislations, as well as being able to store and maintain the knowledge among the personnel. In total, the company sees a general increase in efficiency in several areas.

Based on the implementation work, the company forecasts a decreased design lead-time of at least 10%, probably more. This is mainly due to reducing the number of reworks among the engineers. At the company, there are 60 design engineers that have 108 000 work hours annually. At an hourly rate of 700 SEK, this equals a personnel cost of 75 600 000 SEK. A decreased design lead-time equals an annual cost reduction of 7 560 000 SEK or a decrease in personnel by six persons.

The numbers to use for the payback calculation are as follows:

- Reduced annual design cost: 7 560 000 SEK.
- Investment cost year 1: 604 000 SEK.

These numbers are inserted in the payback table, see **Table 1**, where the investment is made at year 1. In the following years 2, 3, and 4, the company will acquire the gains in design cost. At the end of a 4-year period, the accumulated cash flow is positive results of approximately 22 MSEK.

9. Conclusions

By structuring the design work and planning for the design process, gains will be made in design quality, lead-time, and cost. The design process TWWD has been developed and implemented in several companies for improving the design work. In the implementation case described in this chapter, the investments in improving the product design process will have a major positive effect on the product design years

Payback TWWD implementation	Year 1	Year 2	Year 3	Year 4
Investment cost	−604 800 SEK			
Cost savings in design		7 560 000 SEK	7 560 000 SEK	7 560 000 SEK
Total net cash flow	−604 800 SEK	7 560 000 SEK	7 560 000 SEK	7 560 000 SEK
Accumulated cashflow	−604 800 SEK	6 955 200 SEK	14 515 200 SEK	22 075 200 SEK

Table 1.
Payback calculation for the implementation at the vehicle electronics company.

ahead. Typically, the benefits are gained in lead-time and development cost, which have shown to be rather easy to calculate.

Even though TWWD has shown to be one approach to improving the product design process. There should be more cases of implementations before one can say that the process will generally work for any (many types) company. Also, there is still work to be done to better calculate the improvements in quality.

Finally, one may conclude that there is a need for companies to adopt a structured product design process (such as TWWD). Specifically, since the competitiveness also depends on the actual design process maybe more than ever.

Acknowledgements

My gratitude to all collaborators within the Swedish and Asian industries that have been part of developing and implementing the TWWD process for a period of 15 years. I am also grateful to the research fellows at Dalarna University that has brought useful input to the work. Finally, thanks to all the inspiring discussions with students that have learned the TWWD process during their engineering design education.

Author details

Patrick Kenger
School of Technology and Engineering, Dalarna University, Borlänge, Sweden

*Address all correspondence to: patrik.kenger@gmail.com

References

[1] Norden PV. On the anatomy of development projects. IRE Transactions on Engineering Management. 1960; **EM-7**(1):34-42

[2] Nigel C. A history of design methodology. In: aL MJdVe, editor. Design Methcdology and Relationships with Science. Dordrecht: Kluwer Academic Publishers; 1993

[3] Pugh S. Total Design - Integrated methods for Successful Product Engineering. UK: Addison-Wesley Publishing Company Inc; 1991

[4] Pahl G, Beitz W. Engineering Design - A Systematic Approach. London: Springer Verlag; 1996

[5] Hubka V, Eder WE. Theory of Technical Systems. Berlin: Springer-Verlag; 1988

[6] Andreasen MM, editor. The Theory of Domains. Workshop on Understanding Function and Function to Form Evolution. Cambridge, UK: Cambridge University; 1992

[7] Boothroyd G, Dewhurst P. Product Design for Assembly. Wakefield: Boothroyd Dewhurst, Inc.; 1989

[8] Ulrich KT, Eppinger SD. Product Design and Development. 2nd ed. New York: McGraw-Hill; 2000

[9] Olhager J. Produktionsekonomi (In Swedish). Lund: Studentliteratur; 2000

[10] Baldwin CY, Clark KB. Managing in an Age of Modularity. Boston: Harvard Business Review; 1997

[11] Fixson SK. Modularity and commonality research: Past developments and future opportunities. Concurrent Engineering. 2007;**15**(2): 85-111

[12] Taylor FW. The Principles of Scientific Management. New York: Dover Publications; 1911

[13] Erixon G. Modular Function Deployment - A Method for Product Modularization. Stockholm: the Royal Institute of Technology; 1998

[14] Radford GS. The Control of Quality. New York: Industrial Management; 1917. pp. 100-107

[15] Kenger P. Module Property Verification. Royal Institute of Technology: Universitetsservice US AB; 2006

[16] Simpson TW. Product platform design and customization: status and promise. Artificial Intelligence for Engineering Design, Analysis and Manufacturing. 2004;(18):3-20

[17] Wynn DC, Clarkson PJ. Process models in design and development. Research in Engineering Design. 2018; **29**(2):161-202

[18] Ohno T. Toyota Production System: Beyond Large-Scale Production. New York: Productivity Press; 1988

[19] Womack JP, Jones DT. Lean Thinking : Banish Waste and Create Wealth in Your Corporation. USA: Free Press; 2003

[20] Baudin M. Lean Assembly – The Nuts and Bolts of Making Assembly Operations Flow. New York: Productivity Press; 2002

[21] Womack JP, Jones DT, Roos D. The Machine that Changed the World: Based

on the Massachusetts Institute of Technology 5-Million-Dollar 5-Year Study on the Future of the Automobile. New York: Macmillan Publishing Company; 1990

[22] Morgan JM, Liker JK. The Toyota Product Development System. New York: Taylor & Francis Group; 2006

[23] Kennedy MN. Product Development for the Lean Enterprise. Richmond: The Oaklea Press; 2003

[24] Tahera K, Wynn DC, Earl C, Eckert CM. Testing in the incremental design and development of complex products. Research in Engineering Design. 2018;**30**(2):291-316

[25] Kim K. Lee K-p. Collaborative product design processes of industrial design and engineering design in consumer product companies. Design Studies. 2016;**46**:226-260

[26] Sroufe R, Curkovic S, Montabon F, Melnyk SA. The new product design process and design for environment: "Crossing the chasm". International Journal of Operations & Production Management. 2000;**20**(2):267-291

[27] Wynn DC, Eckert CM, Clarkson PJ. Research into the design and development process: some themes and an overview of the special issue. Research in Engineering Design. 2019; **30**(2):157-160

[28] Barton JA, Love DM, Taylor GD. Design determines 70% of cost? A review of implications for design evaluation. Journal of Engineering Design. 2001;**12**(1):47-58

[29] Ulrich KT, Pearson SA. Does Design Really Determines 80% of the Mancufacturing Cost. Cambridge: Massachusetts Institute of Technology; 1993

[30] Directive 2006/42/EC of the European parliament and on the council, on machinery, and amending Directive 95/16/EC (recast), the official journal of the European union. 2006

[31] Nevins JL, Whitney DE. Concurrent Design of Products and Processes – A Strategy for the Next Generation in Manufacturing. USA: McGraw-Hill; 1989

[32] Andreasen MM, Hein L. Integrated Product Development. United Kingdom: IFS Publications; 1987

[33] Ulrich KT, Eppinger SD. Product Design and Development. 3rd ed. New York: McGraw Hill; 2003

[34] Browning TR, editor. Use of Dependency Structure Matrices for Product Development Cycle Time Reduction. In: Fifth ISPE International Conference on Concurrent Engineering: Research and Applications. Tokyo; 1998

[35] Chen C-H, Ling SF, Chen W. Project scheduling for collaborative product development using DSM. International Journal of Project Management. 2003; **21**(4):291-299

[36] Steward DV. The design structure system: A method for managing the design of complex systems. IEEE Transactions on Engineering Management. 1981;**EM-28**(3):71-74

Chapter 5

Perspective Chapter: Communication-Oriented Automobile Design by a Three-Dimensional Sculpture Educational Model – Narrowing the Gap between Company Work and Education System

Ara Jung and Kang Hyun Lee

Abstract

As the paradigm of the automotive industry is changing, new design attempts are being developed for automotive design in various fields. Design education needs to be changed in response to this rapidly changing automotive industry. Particularly, due to the specificity of automobile design work is progressing, with clay modeling as the center point of the design process it is necessary to provide professional learning about three-dimensional sculpture. However, the current undergraduate curriculum lacks the expertise of 3D sculpture education. Based on the previous studies, related to the design background field to reduce the gap with the business, I proposed an effective method for automobile design, ascribed by a 3D sculpture educational model. First, we proposed a more practical and systematic 3D sculpture educational model that can reduce the gap between practice and education to satisfy the educational model. Therefore, this study is based on the analysis of "automotive 3D sculpture elements" in order to creatively solve various problems Second, we proposed efficient communication methods for a professional designer in the field to learn with the competence of teamwork based on the types of communication networks. Third, we further analyzed the design critique types for evaluation and feedback in class usage and suggested the most effective methods.

Keywords: automotive design, 3D sculpture education, clay modeling, design communication, clay modeling education

1. Introduction

1.1 Background of the study

As the paradigm of the automobile industry is changing, the concept of design of automobiles is expanding from 'mobile means' to 'moving space'. At this point, as new attempts are actively made in various fields such as products, spaces, and services, the scope of automotive design is expanding. Therefore, basic education is more important than anything else for students of various fields who do not have the basic knowledge of automotive design. In particular, due to the physical characteristics of automobiles, 3D sculpture education is an indispensable element. A car is an existential entity that we ride and use directly, and it is not limited to simply seeing with our eyes; therefore, it is essential that the mock-up process in which the designer directly sees and perceives the lines, surfaces, highlights, and volumes that make up the car works while working. However, if you look at the current domestic automotive design education, most of the related undergraduate faculties are adopting the entrance exam method centered on two-dimensional (2D) modeling, undergraduate education also has a curriculum focused on concept sketching, and 2D-oriented classes aiming at creative ideas are being conducted.

In addition, collaborative work between experts in each field is very important in the industrial design field including automotive design. In the case of automotive design, more diverse fields of expertise are required to design and mass-produce innovative automobiles that satisfy complex structures, safety requirements, and customer needs. To this end, companies are showing a trend to subdivide and specialize in each field. However, even after completing four years of undergraduate education in automotive design, students feel limited in realizing and substantiating ideas through collaborative work with experts in each field as an automotive designer in the field.

1.2 The need and purpose of the study

To derive creative results required in the new automotive design paradigm, the need for 3D sculpture education in the undergraduate course for students is growing more. In addition, to reduce the gap between current education and present work and for students to materialize their designs as professional designers and mass-produce them, it is necessary to increase the concentration of communication education for collaborative work.

Thus, the goals of the education model proposed in this study are as follows: First, to esthetically shape the beauty of formativeness in the formative aspect, which is the essence of design, it is necessary for students to understand the 3D modeling of automobiles and, based on it, try to creatively solve problems in 3D modeling. Second, in the 'practical aspect', through communication for collaborative work, the gap between education and work is to be narrowed, and professional designers who create creative modeling are to be nurtured.

2. Automotive design and education

2.1 Changes in automotive design

In the process of development of the automobile industry in the early days, logical and mechanical processes for effective performance of automobile functions were prioritized in automotive design. However, with smart cars and shared cars, future automotive design is developing in various directions of emotional engineering that respects the complex emotions, psychological satisfaction, personal taste, and individuality obtained from the user's personal experience. As a result, many changes were made to the scope of space design, service design, and visual design and the process accordingly.

First, the change in power technology from an internal combustion engine to an electric motor not only plays an important role as an alternative to energy problems and environmental problems but also brings major changes to automotive design. Depending on the location where the vehicle's power is transmitted, the large structure of the vehicle is designed to center the weight of the vehicle with most of the mechanical elements in the front. Because of these mechanical elements that must be included indispensably, changes in the design of automobiles had to be made in small portions. However, as Gu [1] mentioned, as the vehicle's power is changed to electricity, the engine room and passenger room are not separated; their utilization as a single space can be increased; and the scope of structural and morphological changes such as profile and proportion has been greatly expanded.

Second, when autonomous driving becomes possible, automobiles can be expanded into a living space beyond a simple means of transportation. Automotive interiors for driving also had limitations in design changes due to essential elements. Currently, automobile interior design elements are designed with a focus on functions for driving. If autonomous driving is activated in the future, interior design elements such as livability and comfort should be more emphasized in the interior, and not only the arrangement structure of all elements will change but also the components for operation must be removed, and in addition, it is expected that components for user convenience will change, leading to major changes in the design.

Third, as the paradigm of automobiles changes and its meaning expands, companies in various fields such as IT, electric and home appliance companies, Google, Apple, Uber, and Alibaba are also entering the automobile industry. Also, automotive design is being tried in various fields such as engineering major, psychology major, spatial design, and visual design major in addition to the existing industrial design major.

The automotive design process is also expanding the range of choices by diversifying trims to respect the individuality of customers, and with the development of tools and programs for design, it has become possible to design a variety of designs.

As such, even at the present time when automotive design is expanding and rapidly changing, to check the sense of weight, space, and volume due to the characteristics of automotive modeling, still, the clay model plays an important role in the design process.

Contrary to the expectation that all industries will move from offline to online, recently, the O2O (Online to Offline) platform, which combines the advantages of online and offline businesses, has been applied to various fields and is receiving great response from customers.

In automotive design, the design process using digital data is also developing, but the clay model, which can be touched by humans and is a design tool through experience, is still the focal point of the design process, and for more creative design, converting a concept sketch (2D) into three dimensional (3D) is an important starting point in design.

2.2 Current status of domestic automotive design education

In this chapter, we will examine the current status of automotive design education in domestic universities. This study analyzed five major universities in the metropolitan area (Hongik University, Korea National University of Arts, Kookmin University,

Chung-Ang University, and Seoul National University) that implement automotive design education as their major subject.

The automotive design classes for each school are organized in a table so that the subjects and contents of the classes can be viewed by emphasizing the 3D modeling classes by grade and semester. However, in the case of Chung-Ang University and Seoul National University, where detailed class content is not listed on the website, the class content was excluded (**Table 1**).

Academic year	1st semester	2nd semester	Content
	Subject		
1	Free association and expression techniques 1	Free association and expression techniques 2	Creative idea, expression, and delivery
	Basic 3D 1	Basic 3D 2	Basic expression of 3D modeling based on points, lines, and surfaces
2	Transportation design studio 1	Transportation design studio 2	Basic skills training in sketching, rendering, and drawing
3	Practical drawing		Idea development, concept sharing, communication, and rendering
4	Transportation design studio 3	Transportation design studio 4	Transport equipment structural analysis, formative research, and experiment
	Design seminar		Invite experts, broaden your horizons through indirect experience
	portfolio 1	portfolio 2	In-depth course, abstract theory learning through practical design course
	transportation design studio 5	transportation design studio 6	4-year course creative portfolio creation (goals for employment and graduate study)

Color defines the 3D modeling education class intensity among other curriculum's.

Table 1.
Curriculum of industrial Design Department of Hongik University.

2.2.1 Hongik University - department of industrial design

2.2.2 Korea National University of arts – design

See **Table 2**.

Academic year	1st semester	2nd semester	Content
	subject	subject	
1	Basic transport design 1	Basic transport design 2	Basic concept and process case study for transportation device design (new 2017)
	Basic digital design & modeling 1	Basic digital design & modeling 2	Understanding and expressing the 3D design process using Alias
	Basic model shop		Mock-up and prototype for finishing quality of design (using various materials and data)

Academic year	1st semester subject	2nd semester subject	Content
2	Modeling and product 1	Modeling and product 2	A beautiful design that conforms to the formative beauty, functional recognition, emotion, and logic of individual subject products
	3D Printing & modeling 1	3D Printing & modeling 2	3D printing for mock-up (new 2017)
	Digital modeling 1	Digital modeling 2	3D modeling using the Alias program, hyper-realistic rendering, and hard mock-up production using CNC
3	Object studio 1	Object studio 2	Individual items for each student, harmony of modeling/function/sensibility/usability, linkage with other studios, broad perspectives, and results from various perspectives
		Model shop 1	Real mock-up covering various materials and forming techniques and machine tools in the modeling process
4	Object studio 3	Object studio 4	Individual items for each student, harmony of modeling/function/sensibility/usability, linkage with other studios, broad perspectives, and results from various perspectives
	Model shop 2		Modeling process, real mock-up covering a variety of materials, forming techniques, and machine tools

Color defines the 3D modeling education class intensity among other curriculum's.

Table 2.
Curriculum of Design Department of the Korean National University of arts.

2.2.3 Kookmin University - department of automotive transportation design

See **Table 3**.

Academic year	1st semester		2nd semester	
	Subject	Content	Subject	Content
1	Industrial design theory	The concept, principle, and development process of modern design	Introduction to automotive design	A change in the production method and paradigm of transportation equipment
	3D modeling	Basic 3D modeling technique using the concepts of fractals and topology	Basic expression techniques	2D sketch technique training
	Basic drawing	Basic practice of hand drawing	Digital modeling 1	Alias literacy training
	Digital drawing 1	Practice on Photoshop utilization	Basic automotive transportation design studio	Small transport device-oriented design development process
2	Modernism studies	Characteristics, history, and formativeness of modernism	Design methodology	Design results that maintain restraint and esthetic value
	Automotive engineering fundamentals	Understanding the principle of automobiles, discussing the latest	Expression techniques 2	Visual expression techniques for effective delivery

Academic year	1st semester		2nd semester	
	Subject	Content	Subject	Content
		technology issues, problems, and finding countermeasures		
	Lettering	Elegant lines, the ability to express shapes	Digital modeling 2	Training the ability to use the 3D-Alias
	Expression technique 1	Visual egression techniques to effectively communicate to third parties	Automotive transport studio 2	Passenger sedan or minivan basic design course (more than 5 passengers)
	Digital drawing 2	Practice using 2D-Alias Sketch-Pro	Automotive transport studio 2	Passenger sedan or minivan basic design course (more than 5 passengers)
	Automotive transportation studio 1	Basic design process for transportation equipment (focus on 3-wheel and 4-wheel small vehicles)	Maya & 3D printing	Image technique and modeling results using Maya are reviewed through 3D printing process
3	Introduction to emotional design	Design from die emotional perspective, case study and methodology, CMF analysis	Automotive design trend	Implications for the development direction of automobile design according to changes in production methods
	CMF	Effects of physical and psychological properties of materials	Imagination studio	A means of transportation that exists in a virtual face based on stories and imagination
	Expression technique 3	Visual expression using complex materials and software	Expression techniques 4	Visual expression using complex materials and software
	Advanced digital modeling 1	Alias modeling advanced course	Advanced digital modeling 2	Alias modeling advanced course
	Automotive interior design 1	Interface-centric design	Automotive interior design 2	Interface-centric design
4	Automotive design studio 1	Style-focused forts automotive design practice	Automotive design studio 2	Style-focused forts automotive design practice
	Transportation design studio 1	The physical and psychological characteristics of color and its use in understanding design	Transport design studio 2	Understanding the concept of high value-added transportation design through leisure equipment design
	Capstone design 1	From product planning to production, practice in the entire process, practical mind, and experience	Capstone design 2	From product planning to production, practice in the entire process, practical mind, and experience
	Advanced expression techniques 1	Presenting visual images for graduation works in various media	Advanced expression techniques 2	Presenting a creative perfective through comprehensive image work such as transportation means and figures in an imaginary space

Academic year	1st semester		2nd semester	
	Subject	**Content**	**Subject**	**Content**
	portfolio 1	Cultivating the ability to organize creative results and present them to third parties	Portfolio 2	Cultivating the ability to organize creative results and present them to third parties
	Automotive design studio 3	Futuristic vehicle design course practice for graduation work	Automotive design studio 4	Completion of a futuristic vehicle using a variety of power sources for graduation work
	Transportation design studio 3	Practicing future-oriented vehicle design process using various power sources	Transport design studio 4	Practicing future-oriented vehicle design process using various power sources

Color defines the 3D modeling education class intensity among other curriculum's.

Table 3.
Curriculum of Department of Automotive Transportation Design at Kookmin University.

2.2.4 Chung-Ang University- industrial design

See **Table 4**.

Academic year	1st semester subject	2nd semester subject
1	Computer basic design	Color design
	Digital application design	
2	Basic design sketch	Advanced design sketch
	CMF design	Advanced digital application design
3	Transportation and 3D concept design (1)	Transportation and 3D concept design (2)
4	Transport equipment design (1)	Transport equipment design (2)

Color defines the 3D modeling education class intensity among other curriculum's.

Table 4.
Curriculum of industrial Design of Chung-Ang University.

2.2.5 Seoul National University- industrial design major

See **Table 5**.

Academic year	1st semester subject	2nd semester subject
2	Basic model making practice	
3		Transport equipment design
4	Transport equipment design project 1	Transport equipment design project 2

Color defines the 3D modeling education class intensity among other curriculum's.

Table 5.
Curriculum of industrial Design of Seoul National University.

2.3 Limitations of automotive design education

Although it is possible to learn the overall contents necessary for automotive design at Hongik University, instructor's expertise and time may not be enough to cope with various contents including 3D modeling, with class contents focusing only on automobile designers and studio-centered class composition. In addition, there is a limitation in that the part about collaborative work in various fields in automotive design and the class on 3D modeling consist of the content on basic modeling principles that are not related to automotive design.

At Korea National University of Arts, the curriculum that responds quickly to trends and abundant teaching materials are used to present their ideas in digital modeling using Alias software, and classes using 3D printers and CNC are relatively well organized. However, these mock-up classes have a large proportion of digital models, and classes on clay modeling used in the field and theory classes on professional automotive 3D modeling should be reinforced.

The Department of Automotive Transportation Design at Kookmin University has the strength that it consists of various classes on various fields necessary for automotive design. However, from the perspective of 3D sculpture education, there is a limitation in that it consists only of classes using software. Therefore, 3D sculpture education using various methods and materials is required, and in particular, a class on clay modeling, which is the most used in the field, and a class on professional automotive stereoscopic modeling are necessary.

Also, like the goal of "cultivating designers with expertise as an industry member with a practical sense", it is necessary to learn about effective collaborative work between each field required in the present work field.

At Chung-Ang University, you can study general industrial design, but it is difficult to expect professionalism in three-dimensional automotive modeling education in automotive design with classes for two years in the 3rd (junior) and the 4th (senior) years. 3D sculpture education is centered on modeling education using software, and education on collaborative work as well as ideation is required.

The classes at Seoul National University are mainly composed of an integrated course that can cultivate the creativity and esthetic skills that designers should have. However, in the case of automotive design class, mock-up class is for one semester, and automotive design class is for three semesters, so it is difficult to expect expertise in automotive design education. It seems that both the time and the content of classes are insufficient to learn automotive design mock-up education and design project performance ability. A professional class on automotive design is required.

Finally, to summarize the automobile education of each university, the educational goal of each school is to 'nurture professional designers who can mass-produce automobiles that lead the future with their own original ideas in the industry'. Although the classes focused on ideas and concepts are relatively well conducted, for mass production, it seems that class research is needed to narrow the gap between the ability to materialize ideas and practice. In particular, 3D sculpture education for automobiles is focused on modeling education using software. However, for students who lack the basics of 3D modeling, 3D sculpture education in which students can actually see and feel should be preceded in order to develop formativeness, and to this end, theoretical research on 3D modeling for automotive design and education using clay, which is the most used in the field, should be conducted.

3. Theoretical review on automobile 3D sculpture education

In this chapter, we will examine the theories necessary for the educational model proposed in this paper through previous studies and literature review. The importance of a clay model among tools for 3D modeling in automotive design will be examined through previous studies and literature review, and elements of 3D modeling of automobiles will be analyzed. Based on this, we examine theories for creatively solving problems that occur when making three-dimensional sketches. To graft collaborative work into education to reduce the gap between education and the present work, we examine communication network types based on previous studies and suggest effective types. Lastly, to use the design critic tool as an evaluation method to increase the learning effect, based on previous studies, we attempt to suggest the critic types necessary for automotive modeling education.

3.1 The importance of clay models in automotive design

As can be seen in a previous study in **Table 6**, which mentioned the importance of 3D modeling of automobiles for the past 10 years, among the automobile mock-up processes, the clay model has been traditionally used as a tool for design development in the automotive design process.

In the past, although there was a prediction that automotive design would be done only with digital tools, Jang [5] said that making a clay model in which a designer can directly see, touch, and feel the scale of a car is the best way to revise a design and is an important design element included in the whole process of design development. In particular, in the case of scale models, according to the study of Bing [4], it is a stage before the production of a full-size model, and it is said that it is the basis for a new and balanced developmental modeling.

Due to the nature of the material, clay has hard physical properties that can be 'sculpted' at room temperature, and at high temperatures, it acquires soft properties that can be shaped by the 'modeling' technique. Because of these characteristics, clay can use all of the techniques of 'sculpture', so that even if the design is modified, a high-level model can be created in a short time, and it is the most useful material for

Name	Content
Gu [2]	A car is an existing entity and an existential tool. Only when you can physically handle such a tool accurately can you create a vital automotive design that moves people's hearts.
Furioso [3]	It can be said that the role of the modeler who makes clay models in the design field is very important. When expressing strange nuances, it is more efficient to proceed with human hands.
	After all, a car that becomes a commodity is a 3D object; at some point, the shape and presence of a 3D object must be confirmed. This is because, in design, especially styling, beauty, and ugliness cannot be judged by numbers alone.
Bing [4]	A reduced model is a very useful means for 3D verification of an idea because it can implement ideas faster and can be easily modified compared to 1:1 model production.
Jang [5]	To develop a design in a 3D form, using clay is the best way, and it is one of the tasks that must be done in most automotive design.

Table 6.
Previous studies of the importance of clay modeling.

making 3D objects because of its good machinability, such as a milling machine, and is widely used to make automotive models around the world.

3.2 Analysis of elements of 3D modeling in automotive design

In the case of 3D sculpture education, classes are conducted focusing on basic concepts such as 'point, line, and surface' or 'balance, repetition, harmony, and rhyme' at school.

Thus, this study attempted to analyze 3D modeling with a focus on automobile modeling. As mentioned by Jang [5], based on Johannes Itten's division of shapes using circles, triangles, and squares in Bauhaus basic education and the relationship between space and 3D and three-dimensional construction training discussed by Joost Schmidt, we analyzed the 5 basic elements that make up automotive design that include profile, proportion, line, volume, and highlight.

3.2.1 Profile

When doing automotive design, the profile means the outline of the side of the car. In this study, along with the profile, the stance commonly referred to as the posture of a car was included in the profile, and the study was conducted (**Figures 1** and **2**).

Profiles represent the nature of automotive design. In designs such as speedy, stable, muscular, slender, sleek, feminine, masculine, heavy, light, and cute, the part that can express the character of the overall design is the profile. This is a difficult part to express numerically, and it is the part that should be coped with most in the scale model stage before making a 1:1 model.

3.2.2 Proportion

The purpose of this study is to identify which parts are important components of proportion and to understand their relationship. When realizing a sketch, it is important to ask, 'What kind of relationship do respective design elements form with each other to form a unique proportion?' Unlike the plane, which is fixed with one point of view, there is a lot of room for different interpretations depending on the viewer's point of view or angle of vision in a 3D space. To objectify these parts, the discussion should be focused on the design sketch, and at this time, it is important not to look at each element partially in the sketch that is fixed at one point of view but to understand the relationship between each other (**Figure 3**).

3.2.3 Line

This study focused on the tension of the line. The part that is easy to miss when realizing the sketch is the tension of the line. In the sketch basic education, handwriting is taught. When a skilled designer sketches, his/her handwriting naturally comes out, which also serves as a standard for sketching skills. Line tension is expressed in sketches with handwriting, and for instance, the Mickey Mouse character drawn by a professional and the Mickey Mouse character drawn by a non-professional show the biggest difference in its line tension. It can distinguish between 'fake characters' and 'real characters', and the 2D character's vitality and dynamics are also expressed. The tension of the line containing the designer's intention is also expressed in the automotive sketch, and it may indicate strength-weakness, tension, movement, rhythm, and weight of the

DOI: http://dx.doi.org/10.5772/intechopen.110203

Figure 1.
Profile analysis.

Figure 2.
Comparison of stance.

automobile. The tension of the line must be expressed in the clay model so that the designer can express the feeling of the sketch he/she wants (**Figures 4** and 5).

3.2.4 Volume

In this study, volume is a concept in which surface and line are combined in automobiles, and this study tried to explain the relationship between the surface and

Figure 3.
Proportion analysis.

Figure 4.
Tension of character line.

the line when realizing the automotive sketch through the relationship between the structure and the line that make up the space in 3D space (**Figure 6**).

Felice Varini in **Figure 7** is an artist who uses an anamorphic technique to draw a two-dimensional figure in a 3D space. To take these works as an example, at a certain fixed point of view, the artist's intentional geometric pattern drawing is visible, but at different angles, this drawing collapses and scatters, showing meaningless color fragments or lines. This is because all elements (wall, column, floor, ceiling, window, railing, etc.) that composes the space and the lines drawn on it are connected to each other. To take a different approach, even if the viewing angle or viewpoint of the work is fixed, if the elements that make up the space are transformed, it looks as if the lines will appear to collapse again and be scattered. This is the same as the relationship between surface and line in an automobile. They are connected and influence each other. Even if the surface of a small part of the automobile changes in volume, angle, and/or position, the line located on it also shows a changed shape, such as a change of direction or a distorted appearance. Conversely, in order for any line to look as intended by the designer without deformation from multiple angles, in addition to the problem of 'Are you holding the line well?', the composition of the surface forming the line should be considered together. **Figures 7** and **8** expresses the feeling of an automobile differently by changing the volume with little change in the design.

3.2.5 Highlight

In this study, highlight means the flow of light on the surface of the car. In automotive design, highlight is the most detailed element and is affected by even the

Figure 5.
Tension of car line.

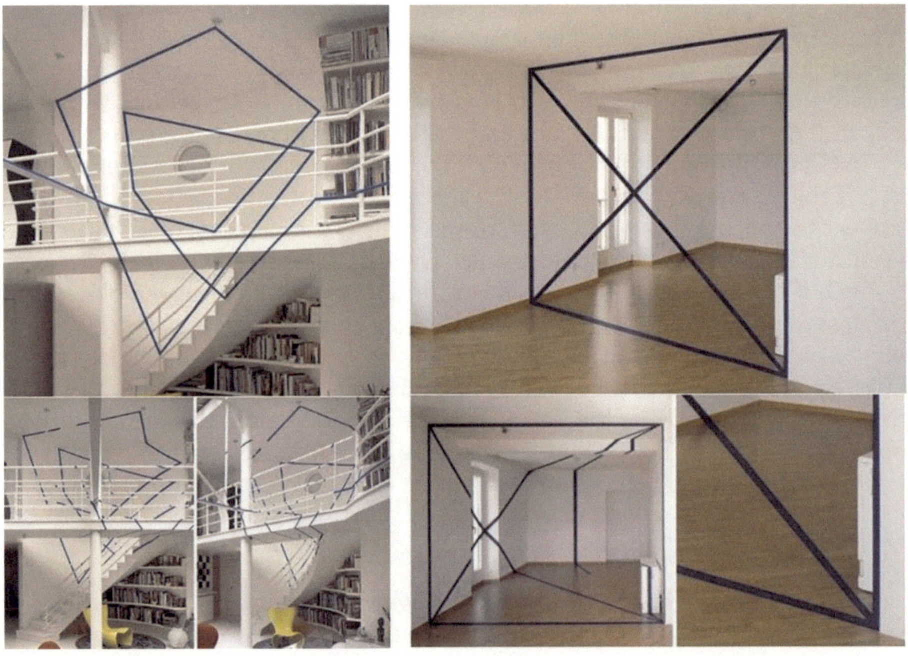

Figure 6.
Anamorphic art.

Figure 7.
Difference of volume.

smallest design change. So, it is most efficient to work after the design is set up, and this is a way to shorten the design time and increase the degree of perfection.

Because the highlight is a very detailed element, even with a difference of 0.01 mm, the flow of the highlight changes, making the car's surface look stiff, dry, or fat, because it has to be controlled with emotional formative elements rather than numerical control; it requires a lot of discussion and experience between designers and clay sculptors. In addition, in the digital model work that transfers this to data, it is necessary to repeatedly check the volume in the clay model, in which very detailed work is required. At this time, as clay has no luster, a special film or foil called Dinoc is attached to the surface to check the flow of light and evaluate. This is the last stage in the design process, and in the final stage of creating data that can make actual molds, highlight is cleaned up through the finest surface adjustment (A Class Surface) process.

3.3 Previous studies on creative problem solving

Creative problem solving means defining and expressing the meaning of things through creative thinking and creating problem solving formatively. The purpose of this study is to find a way to solve the 3D modeling of automotive design and various problems that occur in the process. In the master's thesis of Nam [6], previous studies on creativity are summarized as shown in **Table 7**.

Combining the opinions of various researchers above, first, as humans are born with a creative ability, it is something that can potentially be manifested or developed

Figure 8.
Difference of volume.

by experience or education. Second, it is not to create new things, but to reorganize ideas to classify them in order. In other words, creativity is the ability to re-create existing things anew, and it is an ability that all humans have and can develop. Because creativity is the realization of new combinations, reunions, rearrangements, and relationships from accumulated information, by combining this, when expressing an automotive sketch in a real space as a 3D model, the problem should be solved creatively in the process of three-dimensionalizing a plane by using the 5 elements of 3D automotive modeling mentioned above as 'accumulated information'.

In particular, since it affects the X, Y, and Z axes even if the line is lowered by 1 mm in the solid, in order to obtain the desired result, it is essential to solve creative problems based on the analysis of the elements of 3D modeling.

3.4 Types of communication for design collaborative work

Communication is an essential element in our lives and plays a very important role within an organization. Barnard, an American public administration scholar, argued

Name	Content
Torrance [7]	The ability to find and solve problems; that is, it is the process of detecting difficulties and problems, finding gaps in information, guessing about defects, forming hypotheses, evaluating, verifying, re-correcting, and re-verifying the results to inform the results.
Sternberg and Grigorenko [8]	Thinking is divided into two modes, namely, convergent thinking and divergent thinking.
Sternberg and Grigorenko [8]	Creative thinking is viewed as fluency, flexibility, originality, sophistication, sensitivity, redefining, and restructuring.
Noppe and Gallagher [9]	It is the ability to generate a variety of ideas, where there is a problem or need an idea.
Noppe and Gallagher [9]	It is to create a variety of ideas, come up with original ideas, and develop ideas elaborately.
Lee Seong-jin [10]	It is the ability (creativity) to create a thing or idea of value (utility) by extracting (dismantling) existing information (experience and knowledge) and assembling (combining) a new one.
Lee Seong-jin [10]	To solve a problem, from something to something. By deconstructing the current knowledge and experience, a new combination is created.
Gardner [11]	Seeing creativity as a psychological process. It is defined as an attitude and passion for work, such as perseverance or achievement.
Gardner [11]	It is seen that attitude and passion lead to productive and creative thinking.
Olsen [12]	The inner power that emerges from the individual's uniqueness, yielding valuable new ideas or fresh insights.
Olsen [12]	Creativity is the creation of valuable and new ideas that originate within a person (creativity at a private level).
Youn Jong-Gun [13]	According to a given situation and problem, a new relationship is formed by constructing a new alternative based on past experiences and transferring the new situation.
Youn Jong-Gun [13]	It expands the scope from the past designs that focused on capturing esthetic and emotional values to solving users' problems.
Osborn [14]	To propose a six-stage creative problem-solving model including finding purpose, finding data, finding problems, finding ideas, finding solutions, and finding acceptance.
Osborn [14]	To emphasize the appropriateness of creative problem-solving to equally spend time in each stage on balancing divergent and convergence processes.

Table 7.
Previous studies of creativity.

that goals, cooperation intentions, and communication are the 3 major elements of an organization and emphasized that they occupy an important place in the system that creates an organization. Also, Major and Frone [15] mentioned the importance of organization members by saying that they spend 80% of their daily life collecting information or communicating within the organization.

Summarizing the importance of communication among members, communication contributes to innovative corporate efficiency and creative organizational performance. As can be seen in Kookmin University's talent profile, "Cultivating designers who value teamwork", teamwork is an essential element for mass production design of automobiles, and 'Does collaborative work with members go smoothly?' takes a large part in 'successful design' and 'increase in design efficiency'.

However, as mentioned by Lee [16] and Bing [4] in the field of current automobile education, it emphasizes the importance of 'the designer's individual experience' rather than 'collaborative work using communication'. However, in line with the trend of companies that are segmenting and specializing in the field of automotive design, communication education for collaborative work is necessary so that students who aim to become future designers can demonstrate their competence as a member of a company.

Based on Kim Yeong-im's study [17], this study organized the 4 general communication network patterns such as cartwheel type, chain type, circle type, and interconnection type, and a total of 5 network pattern types including the Y type pattern mentioned in Jeong et al. [18] as shown in **Table 8**.

These patterns can be broadly classified into 2 categories considering the characteristics and efficiency of individual and group work. First, as effective types for individual work, there are cartwheel type, Y type, and chain type, which contain mechanical, dictatorial, centralized, and vertical characteristics. Second, as types that are effective for group work, the circle type and interconnection type, which have organic, democratic, decentralized, and laissez-faire characteristics, fall into this category.

In the book of Jeong et al. [18], the efficiency differences of the 5 network types are summarized in **Table 9**. Based on previous studies on these communication types, if effective types are selected and educated according to the characteristics of automotive design work, it can be a practical education that can bridge the gap between education and practice.

In the case of automotive design work, it is characterized by having both the 'characteristics of individual work' of a designer and the 'characteristics of group work' in which experts in each field collaborate. For this, as information transfer is concentrated on the group leader, there is a strength, which guarantees speed in accurately identifying and resolving problem situations, and it is valid when the nature of the problem is routine and simple, but in the case of complex and difficult

Cartwheel type	Y type	Chain type	Circle type	Interconnection type

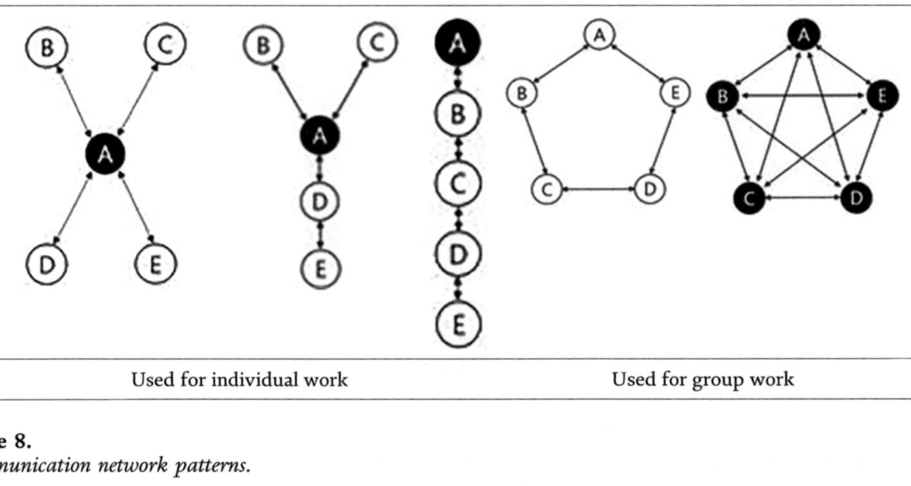

Used for individual work	Used for group work

Table 8.
Communication network patterns.

Types / Efficiency	Cartwheel type	Y type	Chain type		Circle type	Interconnection type
			vertical	horizontal		
Problem-solving speed	fast	fast	fast	slow	slow	slow
Simple problem (accuracy)	high	high	low	low	low	high
Complex problem (accuracy)	low	low	low	low	high	high
Satisfaction	low	low	low	low	high	high
Central figure	A	A	A	C	none	all

Table 9.
Communication pattern efficiency.

problems, the 'cartwheel type', which has the disadvantage that it is difficult to share information directly among members, and because all members exchange necessary information regardless of their team or position, although the actual time required for comprehensive situational awareness and problem resolution gets relatively longer, accurate situation judgment is possible, and it is most effective for difficult and complex problems or problems that require creativity of members, and an appropriate combination of 'interconnection type', which shows the highest level of satisfaction among communication members, is required.

In particular, the 'cartwheel type' communication model, in which the designer is the center, is effective in the prior design stage, which includes the production of scale models. In the scale model stage, the sketch is shaped in the actual space, and if the designer becomes the focal point and communicates with the members, the initial idea can be effectively applied to the model.

In the mass production stage, the 'interconnection type' in which experts in each field communicate directly in complex problem situations is effective. At this time, if you participate in a selective field according to the content or nature of the problem, you can compensate for the slow problem solving speed.

In the educational model of this study, communication practice takes place when producing a scale model based on communication theory learning for collaborative work. Therefore, it is limited to communication practice between 'designer' and 'clay sculptor' based on the cartwheel type as shown in **Table 10**.

The scale model and communication practice are performed twice in total as a team of 2 people, and at this time, they take on the role of a 'designer' and a 'clay sculptor', respectively, and conduct practice. At this time, to materialize his/her abstract ideas and concepts, the student, who plays the role of a designer, will learn specific and objective communication activities that enable communication though persuading the clay sculptor to be understood. For the student playing the role of a clay sculptor, based on the previously learned automotive design modeling elements and creative problem-solving in modeling, the communication activity practice necessary to analyze the designer's sketch more objectively and effectively materialize the sketch is conducted.

3.5 Design critic to increase learning effect

Carson (1994) defined critic as the basic communication behavior of design education. According to Crookes (1994), in general, in the case of a project-type design class,

Communication network types for automotive design collaborative work		
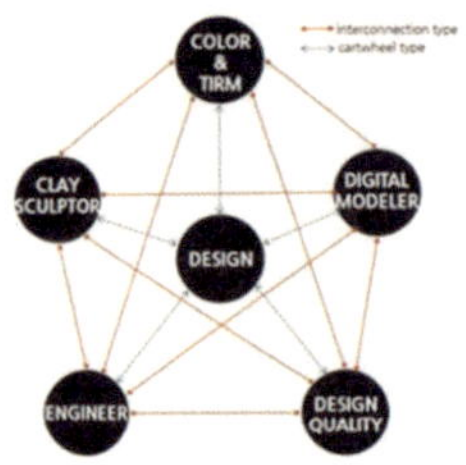	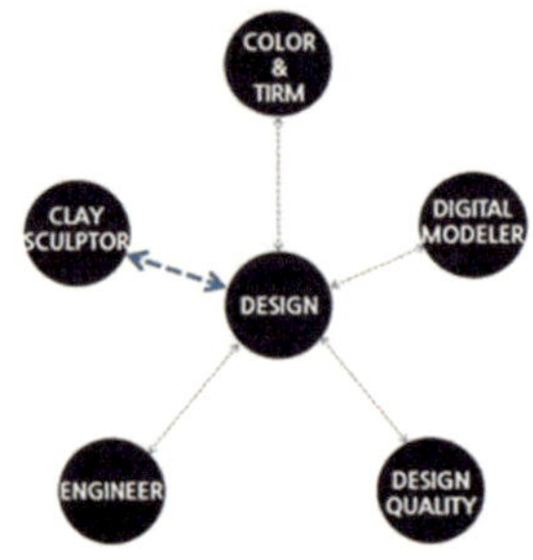	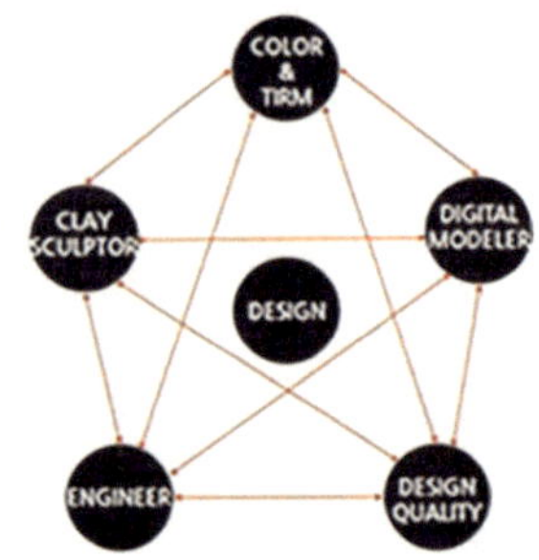
Advance + mass production design stage	Advance design stage	Mass production design stage

Table 10.
Communication styles of automotive design collaboration.

the instructor solves the design problem, the instructor guides students to solve design problems and improve design quality, and not only the instructor but also colleagues participate in class evaluation and discussion. Therefore, students need to make an effort to persuade their design to instructors and colleagues in class. Dannels [19] referred to the dialog form of the interaction that occurs here as a critic and mentioned that it is based on feedback. According to Lu and Bol [20], in various previous studies, it is said that critic not only contributes to the improvement of design quality but also has a learning effect that can help participants generate new ideas.

Design critic types to be used in the proposed 3D modeling model in this study are based on **Figure 9**, which summarizes the design education critic types of Aan [21].

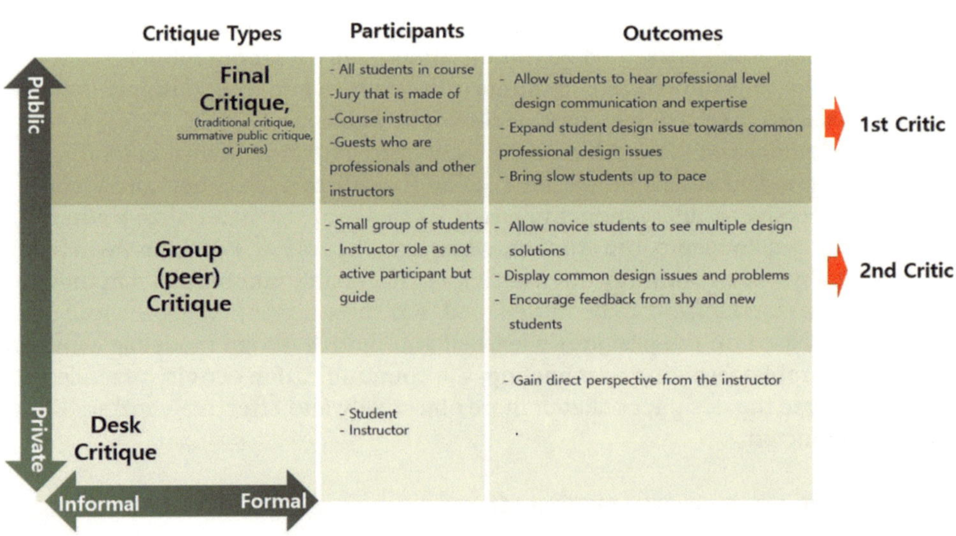

Figure 9.
Types of critique.

In the figure, the characteristics of an informal and formal critique are shown based on the X axis, and the characteristics of a public and private critic are shown based on the Y axis, based on the types presenting the critic's opinion.

Because the effect that can be obtained in class may vary depending on the type of critic, it is important which critic types the instructor chooses in the class. Therefore, appropriate evaluation and feedback can be achieved only when various types of critics are conducted even in one class according to the topic or purpose.

Therefore, this study adopted 2types, namely, 'final critic' and 'group critic' to increase the learning effect of a '3D sculpture education model of automotive design'.

In the case of the 'first critic' in the education model, the critic proceeds with the 'final critic' type as the evaluation and feedback stage in the preliminary practice stage for this practice. In the case of the instructor, when students conduct critic as a stage to help and confirm the theory learned, as an expert, he/she conducts a professional critic of the students' work and expands and organizes the problems presented by the students into specialized parts. Also, he/she plays the role of an assistant for students with retarded understanding. The critic at this stage has a higher proportion of 'feedback' than 'evaluation' of the result, and in this stage, the feedback received through the critic can be supplemented and applied in this practice.

The 'second critic' is the critic stage for evaluation and feedback of this practice. The instructor plays a role in helping the critic proceed, and the students participating in the class are the subject of critics and conduct evaluation and feedback based on the purpose and intention of the work. At this stage, the purpose of the critic is to evaluate the results and share and discuss the feelings and experiences of the students who participated in the class to find solutions to the problems that occurred at this time, and by indirectly experiencing the situation of peer students, it is intended to increase the learning effect by broadening the scope of experience.

3.6 The structure of the 3D sculpture education model of automotive design

The purpose of this chapter is to make a framework of the 3D sculpture education model of automotive design for the realization (3D) of ideas (2D) that are effective for learners.

To establish a framework suitable for '3D sculpture education of automotive design,' which this study intends to propose, of the educational model theories proposed by various scholars in the past, **Figure 10**, an I-P-O model compatible with the characteristics of 3D modeling of automotive design, was adopted as the basic model for this study.

As shown in **Figure 10**, Teiner (1972), McGrath (1984), and Hackman (1987) proposed a group innovation model consisting of 3 stages, namely, input, process, and output, which influence outcome innovation by realizing new ideas in organizational and social activities.

Landy (2016) and Conte (2016) stated that this model is a method of maximizing the efficiency and performance of teamwork and is similar to the framework of the

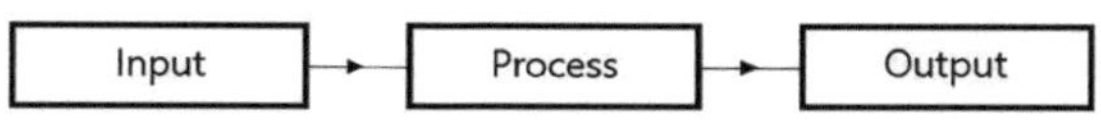

Figure 10.
I-P-O model.

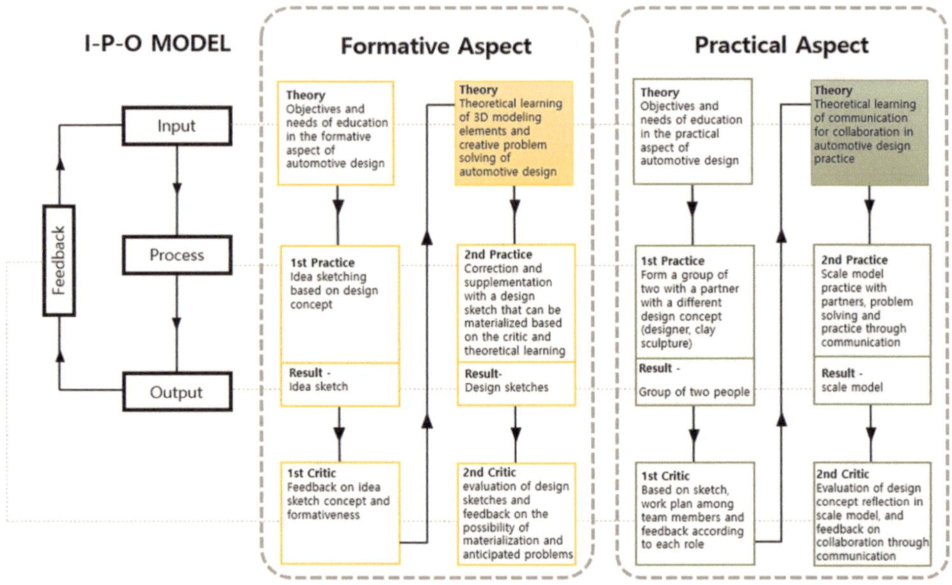

Figure 11.
Automobile design in terms of three-dimensional sculpture educational model framework.

design education process based on the aspects of collaboration, innovation, productivity, and systematic attributes in particular. However, as there is a limitation in the classical I-P-O model, which has a single-cycle path from input to output, based on a framework that repeatedly implements the interaction of input, process, output, and feedback, as shown in **Figure 11**, we readjusted the model by combining the characteristics of 'formative aspect' and 'practical aspect' for 3D sculpture education of automotive design.

3.7 The proposal for the 3D sculpture education model of automotive design

The automotive design education model was studied with a focus on model production (scale model) among the preceding design stages, and it has the structure of the I-P-O model and cycles through the stages of the input, process, output, and feedback.

This study, aiming for students to materialize their ideas as professional designers and to mass-produce automobiles, proposes **Figure 12**, an educational model that combines the 'formative aspect' and the 'practical aspect' to promote the specialization of automotive 3D sculpture education and to reduce the gap between present work and education.

This model is composed of theory, practice, and critic, with the 'formative aspect' and 'practical aspect' proceeding sequentially in the direction of the arrow.

The detailed contents of education are again divided into **Table 11**, which is a 'formative aspect', and **Table 12**, which is a 'practical aspect,' and summarized as follows.

Based on the basic education model, detailed class contents are subject to change according to members, environment, student level, etc.

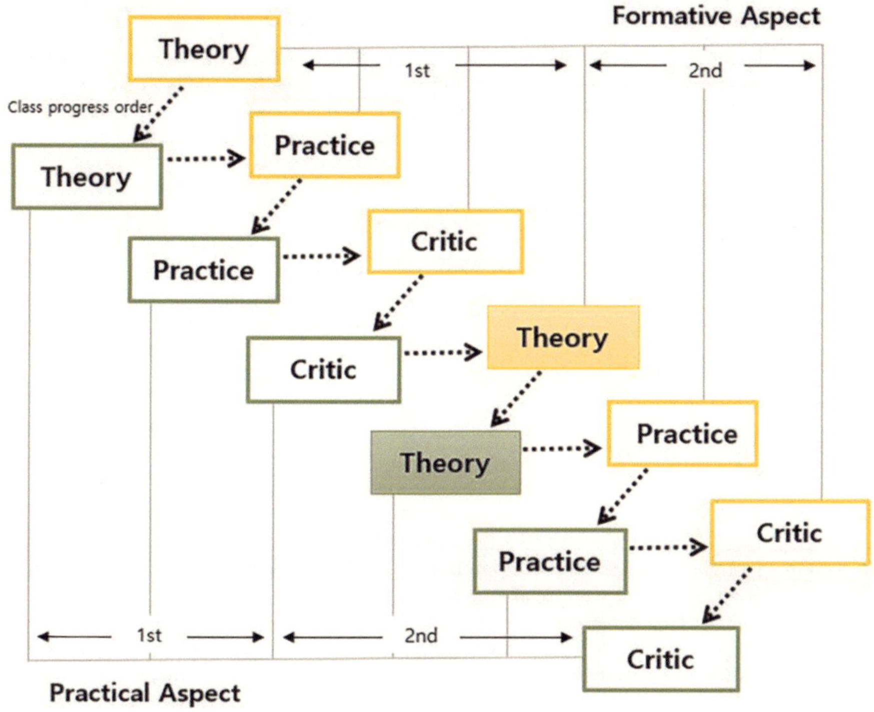

Figure 12.
Automobile design by three-dimensional sculpture educational model.

3.7.1 Detailed model in the formative aspect

	Course	Content	Goal	Details	Remarks
1st	Theory	Education background and needs goal setting	In the formative aspect, understand the background and necessity of this education and set the goal of the class.	Changes in automotive paradigm and changes in automotive design according to industrial development are examined. Accordingly, the importance of 3D modeling in automotive design increases as it expands to various fields of automotive design and enables more free and diverse modeling designs.	
	Practice	Idea sketch in progress	One idea sketch for each student is completed, centered on the design concept.	Prior to sketching, the concept desired by the student (designer) is arranged into 3 to 5 adjectives, and then the idea sketch is proceeded based on this. The materials of the sketch are free, and work is focused on expressing one's own concept rather than quality.	Result—idea sketch

	Course	Content	Goal	Details	Remarks
	Critic	Idea sketch critic	Evaluate whether the concept in the idea sketch is properly expressed in the formative in the idea sketch and find parts to be corrected and supplemented	The instructor's professional and appropriate critic and the students' free critic are conducted simultaneously. Based on the selected adjectives, it is evaluated whether the concept is properly expressed in a formative way in the idea sketch, and feedback is provided on the parts to be corrected and supplemented. At this time, rather than evaluating the results, critics are conducted focusing on feedback on the direction of development so that they can be corrected and supplemented.	final critic types
2nd	Theory	Learning about automotive 3D modeling	Find out how to creatively solve problems in three-dimensionalization by knowing and applying 3D elements of automobiles	Theoretical education is conducted on the five automotive 3D modeling elements such as profile, proportion, line, volume, and highlight, and based on this, the class will try to find ways to creatively solve problems that may occur when converting actual 2D into 3D.	Automotive 3D modeling Creative problem-solving
	Practice	Design sketch in progress	Revision and supplementation with design sketches that can materialize idea sketches based on the first critic content and theory learning	By referring to the content of the first critic, the class will supplement the idea sketch with a focus on the student's concept and think about the problems when making an idea sketch (2D) into reality (3D) based on the knowledge of 3D automotive modeling and proceed with an objective design sketch that everyone can understand.	Result—design sketch
	Critic	Design sketch critic	Criticize the expected problems when materializing the design sketch	The instructor acts as a facilitator, and the students freely conduct the critics. Based on the theory of 3D automotive modeling, revised and supplemented design sketches are evaluated, and feedback is provided for deficiencies. At this time, feedback on unresolved problems is carried out along with evaluation of the clarity and accuracy of the sketch.	Group critic types

Table 11.
Three-dimensional sculpture-related detail model.

3.7.2 Detailed model in the practical aspect

	Course	Content	Goal	Details	Remarks
1st	Theory	Educational background and needs, goal setting	To know the background and necessity of this education from the practical aspect and set the goal of the class	Based on the description of the role of each department and the importance of collaborative work according to the change in the automotive design process, the importance of communication in the field is known, and learning goals are set.	
	Practice	Group formation	To form a group with two people and set up a work plan	To practice communication for collaborative work while carrying out the scale model, students will take the roles of 'designer' and 'clay sculptor' to form a team of two people. At this time, it is effective to practice communication for visual expression of various adjectives by forming a team of students with different design concepts and topics.	Output group of two people
	Critic	Group critic	Based on work plans and sketches with team members, feedback from the perspective of each field	The instructor's professional and appropriate critic and the students' free critic are conducted simultaneously. First, the work order and method are planned through communication with the team members, and a pre-meeting on the design concept is conducted. At this time, criticize freely based on feedback on the necessary parts from the perspective of the role.	Final critic types
2nd	Theory	Learning about organizational communication	Understanding organizational communication for collaborative work in practice and applying it to practice	To understand the concept and importance of organizational communication, learn appropriate types for inter-departmental collaborative work in automotive design among communication network	Communication types for collaborative work

Course	Content	Goal	Details	Remarks
			types, and conduct theoretical learning on verbal and non-verbal (sketch) communication.	
Practice	Scale model practice in progress	Communication practice through the creation of scale models centered on collaborative work with team members	Based on the understanding of automobile 3D modeling, design sketches are produced as scale models in their respective roles. At this time, the selected adjectives and concepts should be formatively reflected in the model.	Output-scale model
Critic	Critics on a scale model	Feedback on evaluation of the scale model and communication with partners	The instructor acts as a facilitator, and the students freely conduct the critics. It is evaluated whether the adjectives and concepts of the design sketch are well reflected in the model based on the theory of automotive 3D modeling rather than the completeness of the scale model itself. At the same time, successful and unsuccessful cases of collaborative work through communication between team members to expand experience are presented, and feedback is provided.	Group critic types

Table 12.
Company work related detail model.

4. Conclusion and a development strategy

The goal of this study was to suggest professional and practical automotive 3D sculpture education by narrowing the gap between education and actual work, which is in line with the goal of the department of automotive design of domestic universities, "cultivating designers with expertise as a member of industry with a practical sense" in order for students to effectively materialize (3D) sketches (2D) in the field. The characteristics of the education model of this study are as follows:

First, theoretical research was conducted on creative problem-solving through analysis of 5 elements for professional education on 3D automotive modeling in the 'formative aspect'.

Second, in the 'practical aspect', communication types for collaborative work, which are the most important parts in practice, were organized, and efficient types were proposed for automotive design work.

Third, design critic was used in the 'educational aspect', and critic types were adopted stage by stage to increase the educational effect stage by stage.

Fourth, in order to propose an education model with a 3D structure by synthesizing the 'formative aspect', the 'practical aspect', and the 'educational aspect', an automotive design 3D modeling model based on a circular shape by adding 'feedback' to the I-P-O model was proposed.

This study is meaningful in that the proposed education model presented a practical and efficient framework for automotive 3D modeling education in line with the rapidly changing industry by synthesizing various aspects of problems. In addition, the 3D education method can enhance the learning effect of students, and it is expected to be used in automotive 3D sculpture education for various classes of people in various fields in the future.

As a future research project, education models should be continuously and specifically developed to provide flexible and efficient education based on the proposed education model of this study. In particular, since this study was targeted at students who received basic automotive design concepts and sketch education, it seems that educational research on automotive design learning is necessary before 3D modeling for students in other fields who lack the basics.

Lastly, the efficiency of this educational model should be verified in various ways.

Author details

Ara Jung[1*] and Kang Hyun Lee[2]

1 General Motors Technical Center Korea, South Korea

2 Hongik University, South Korea

*Address all correspondence to: ara.jung87@gmail.com

References

[1] Gu S. An observation on the changing Facrors on automotive design in 21t. Century. Korea Society of Basic Design & Art. 2010;**11**(4):3-9

[2] Gu S. The Textbook of Automotive Design. Seoul: Ahn Graphics; 2016

[3] Furuso H. Motor fan. Seoul: Golden Bell; 2013

[4] Bing Y. Car design process. Auto Journal. 2008;**30**(5):36-38

[5] Jang E. Research about vehicle design learning program. Korea Science & Art form. 2007;**3**:81-91

[6] Nam S. Research on Design Education Method for Advancement of Creativity. (Unpublished master's thesis). Pohang, Korea: Handong Global University; 2010

[7] Torrance EP. Predicting the creativity of elementary school children (1958-80) —and the teacher "who made a difference". Gifted Child Quarterly. 1981;**25**(2):55-62

[8] Sternberg RJ, Grigorenko EL. Guilfords structure of intellect model and model of creativity: Contributions and limitations. Creativity Research Journal. 2001;**13**(3-4):309-316

[9] Noppe LD, Gallagher JM. A cognitive style approach to creative thought. Journal of Personality Assessment. 1977; **41**(1):85-90

[10] Seong-jin L. Educational Psychology: Its defining characteristics and future perspectives. Korean Educational Psychology Association. 1996;**10**(1):25-43

[11] Gardner H. Creativity: An interdisciplinary perspective. Creativity Research Journal. 1988;**1**(1):8-26

[12] Olsen T. Silences. New York: Feminist Press at CUNY; 2003

[13] Jong-Gun Y. School principals leadership and creativity. The Journal of Creativity Education. 1998;**2**:1-25

[14] Osborn A. Applied Imagination. New York: Scribner's; 1953

[15] Major B, Frone M. Communication quality and job satisfaction among managerial nurses. Group & Organization Studies. 1988;**13**(3): 332-347

[16] Lee K. A study on the experiential creativity in digital design process. Basic Formative Science Research. 2004;**5**(1): 73-83. 11 pages

[17] Kim Y, Jeon B. Communication and Humanity. Seoul: Korea National open University Press; 2004

[18] Jeong D, Song K, Ryu T, Park J. Organizational Behavior-Second Edition. Seoul: Samyoungsa; 2008

[19] Dannels D, Housley Gaffney A, Martin K. Students talk about the climate of feedback interventions in the critique. Communication Education. 2011;**60**(1): 95-114

[20] Lu R, Bol L. A comparison of anonymous versus identifiable e-peer review on college student writing performance and the extent of critical feedback. Journal of Interactive Online Learning. 2007;**6**(2):100-115

[21] Aan J. Korean university Students attitudes towards anonymous peer critiques in graphic design education. Korea Science & Art form. 2015;**22**: 207-220

Chapter 6

Extrusion of High-Resolution Polymer Structures with Direct Ink Writing Techniques

Farnaz Rezaei and Stefan Johansson

Abstract

This chapter explores 3D printing, often called additive manufacturing, with a focus on printing structures with high-detail resolution. There are mainly two techniques that allow for 3D printing in the μm to sub-μm range: two-photon printing and direct ink writing (DIW). The two-photon technique is briefly explained, while the focus here is on DIW since this technique gives an opportunity to print a wide variety of materials. To exemplify high-detail resolution polymer 3D printing, biocompatible cellulose acetate (CA) is selected. Printability and the possibility of printing μm feature-size structures with inks containing different amounts and molecular weights of cellulose acetate are presented. Results indicate that by optimizing inks and printing parameters such as the internal and external diameter of the nozzle, strands down to sub-μm can be printed with high placement control. Various challenges as clogging and low printing speed are also discussed.

Keywords: additive manufacturing, 3D printing, direct ink writing, cellulose acetate, extrusion based printing

1. Introduction

Three-dimensional (3D) structures can be built through layer-by-layer deposition of materials on a suitable substrate. This method called 3D printing or additive manufacturing (AM) allows the fabrication of complex 3D structures from computer-aided design models [1]. 3D printing has several advantages in comparison with conventional manufacturing methods. Some of the important benefits of 3D printing are the freedom of design, shorter lead times, and less waste of material [2]. Creating structures with smaller dimensions, complicated geometries, and much higher resolution is also possible by 3D printing.

The present high-detail resolution AM technologies can be divided into two main categories: (1) light-based 3D printing and (2) ink-based 3D printing. Two-photon polymerization (2PP), the light-based printing technique with true sub-μm detail resolution, has many advantages [3]. Some basic facts regarding this technique will be given for comparison reasons. If we focus on high-detail resolution ink-based techniques, these can be further subcategorized with respect to, for instance, the particular mechanisms for how the ink is extruded from the nozzle [2]. The simple direct ink writing (DIW) technique has become widely interesting for diverse applications due to reasons such as the availability of various materials, the detail resolution, and the scalability [4, 5]. Typical envisioned applications are additive manufacturing of membranes for, for example, waste water, medical devices, and bio-separation. In polymer DIW, precise control over the positioning of the printed fibers is key, and by extruding extremely fine fibers, it is possible to reach μm or even sub-μm detail resolution. In this chapter, we will examine examples of DIW structures using ink containing cellulose acetate, which is the naturally derived polymer [6, 7], and in particular discuss methods to create high-detail resolution structures.

2. Light-based printing

2.1 Method

Two-photon polymerization (2PP) is one of the more powerful light-based printing techniques that enables 3D printing of complex structures with high resolution [8]. A 2PP printer is similar to a stereolithographic printer, but it typically uses a femtosecond laser and a high-resolution specimen table in combination with galvomirrors to scan the laser beam fast [9]. From a printing point of view, the major difference is that it is possible to print at a point inside the photocurable resin due to the so-called two-photon process. The high laser intensity in the focal spot allows two low-energy photons to simultaneous activate local photopolymerization, while there is no photopolymerization outside the focal spot since the photon energy from single electrons is not sufficient there [10]. With a focal spot that has the necessary intensity within a fraction of a μm, it is possible to reach sub-μm detail resolution [11].

2.2 Printing parameters

It is essential to realize the importance of the printing parameters that affect the polymerization process. This involves exploring the relationships among, for instance, the resin formulation, scanning speed, and laser power [12]. Depending on printing equipment and printing strategy, the print speed will vary, but as all other high-resolution techniques, it could take hours to print solid object of mm size.

2.3 Printed structures

We have printed structure with holes ranging from 4 μm to 250 nm to exemplify what can be done with a reasonable amount of tuning. As shown in **Figure 1**, it is possible to print holes down to 500 nm with the 2PP technique, and with some tuning, it is possible to go somewhat further. **Figure 2(a,b)** illustrates a successfully printed complex structure comprising two nozzles with 4 μm diameter in the nozzle opening.

The example in **Figure 3** demonstrates the ability to print free-standing structures within a broad size range, from micrometers to nanometers, and in this case, the finest pin diameter is 0.5 μm. This demonstrates that the 2PP printing technology is suitable to produce structures with detail resolution down to sub-μm, and the examples in **Figures 1–3** suggest that it is possible to design structures for diverse applications.

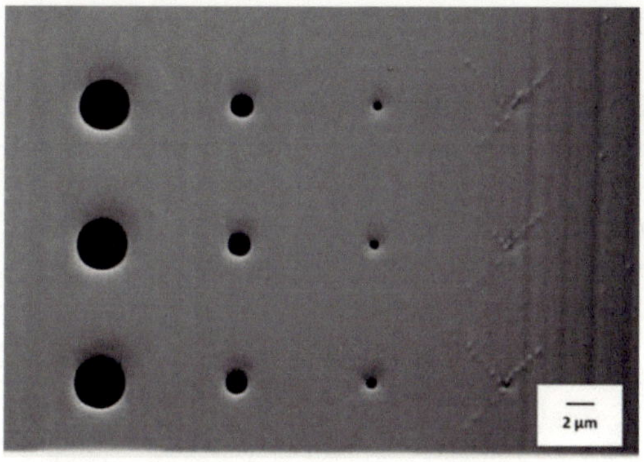

Figure 1.
Printed structure with holes ranging from 4 μm to 500 nm.

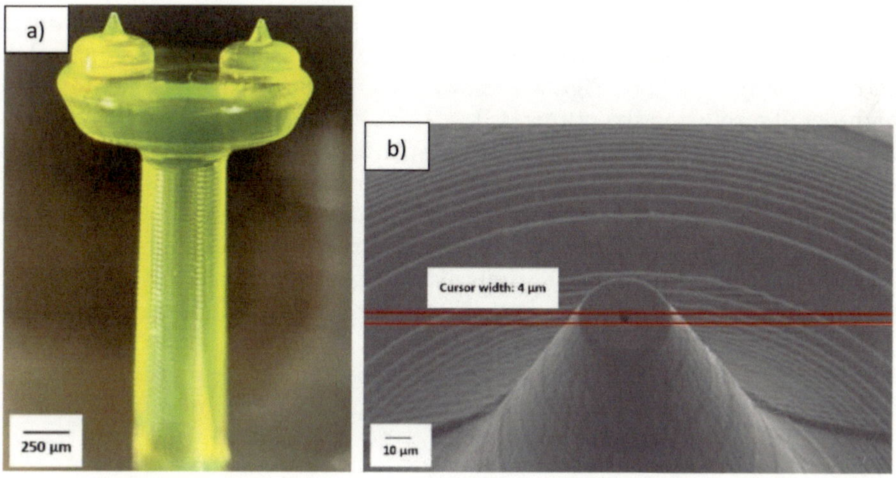

Figure 2.
(a) Two-nozzle printed structure, (b) opening part of the nozzle with 4 μm diameter.

Figure 3.
Printed structure showing pins of various lengths with diameters from 4 to 0.5 μm.

2.4 Benefits and limitations

Due to the possibility to print truly complex structures, for instance, with overhangs, from computer-generated models, combined with the high-detail resolution in several commercially available printers, 2PP has attracted wide attention [13]. However, one of the main factors that limited the progress of 2PP is the challenge of producing components from a broader range of the materials [8]. It should be mentioned that still there are much to explore in the 2PP technique, to fully utilize the capabilities.

3. Direct ink writing

3.1 Method

In the DIW technique, which is an extrusion-based method, ink is deposited through a nozzle on a substrate. After extruding the liquid ink, different parameters such as extrusion rate, print speed, solvent evaporation, and absorption control the solidification of the ink [14]. **Figure 4** shows the printing setup designed for fabricating high-detail resolution structures [15].

The setup includes different components such as a syringe pump (motor to drive the syringe piston, the syringe pump uses a LEGS motor from Piezomotor allowing nm movement control), the head part holding the motor that has a force cell to measure force between head and piston, a syringe that should be filled with the desired ink, a glass capillary with nozzle (various nozzles can be attached to the syringe), and a 3D table that moves the substrate [16]. The 3D table should allow for the desired positioning control, and for high-detail resolution, we usually demand sub-μm repeatability. Not shown in **Figure 4** is a microscope that makes it possible to both observe the printing and place the nozzle of the capillary with the optimal spacing,

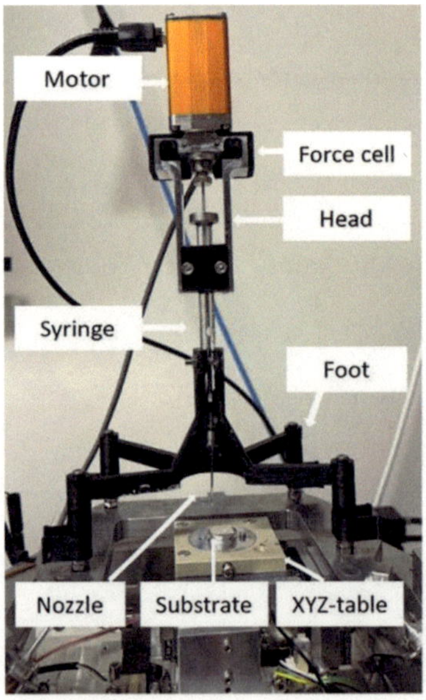

Figure 4.
A direct ink writing setup for high-detail resolution 3D printing.

often in the μm range, from the substrate. The capillaries are typically made long and slender to handle occasional contact with the substrate without braking. The choice of slenderness is delicate since a too slender capillary will result in reduced placement control due to various viscous and friction forces during printing.

3.2 Printing parameters

In the direct ink writing technique, it is important to evaluate the parameters that impact the printing resolution. One of the critical factors is the nozzle size, which can be adjusted based on the desired resolution and application requirements. However, it should be noticed that a reduction of the nozzle size may increase the risk of nozzle clogging. Moreover, shear thinning behavior of the ink in DIW plays an important role in achieving high-resolution structures since this allows for extrusion through narrower nozzles and faster solidification after extrusion. The viscosity of the ink will affect the printing resolution as excessive viscosity hinders the ink extrusion, while insufficient viscosity could cause ink spreading that reduces the lateral resolution of the structure [17].

Printing speed is another parameter that influences the final resolution of the printed structure [6]. Printing speed is typically controlled by the movement speed of the substrate in the X and Y directions. Higher speeds present challenges in the resolution, particularly when there will be high accelerations or retardations as in the corners. It is theoretically possible to adjust the ink flow to compensate for speed changes, but this is a great challenge when ink flow gets close to pl./s range. This flow range corresponds to nozzle diameters in the micron range and speeds of few mm/s, which is often used to get reasonable printing times for micrometer detail resolution structures. Additionally, the applied pressure in the syringe pump could impact the printing results. The compression of the ink and/or gas bubbles will make it difficult to adjust the ink flow, particularly when larger volume inks are under pressure. The pressure should therefore be carefully calibrated in relation to printing speed and ink viscosity. For instance, high pressure could lead to an increased extrusion rate for an ink with low viscosity at low printing speeds, which compromises detail resolution.

3.3 Materials

The DIW method gives flexibility in producing structures for various applications due to the ability of using diverse printable materials. In this chapter, we will focus on polymers since this best illustrates the possibilities to extrude high-detail resolution 3D structures. Among different polymers that can be used in 3D printing, cellulose, which constitutes the main part of a plant's cell walls, has attracted attention due to several interesting properties. Cellulose is biocompatible, and it can be used in various applications like biomedical, pharmaceutical, construction, packaging, clothing, thermal insulation, and so forth. Further, cellulose is a mechanically robust as well as chemically versatile material. Cellulose is insoluble in many solvents; to increase the solubility, derivatives of cellulose can be used. Here, Cellulose acetate (CA) has been selected for its high solubility [18].

3.4 Printed structures

To exemplify the possibilities with direct ink writing, various membrane structures will be discussed. Typically, membrane structures are characterized in terms of physical properties such as pore diameter, pore size distribution, porosity, and tortuosity, which all are dependent on the fabrication method. The placement of the printed strands as well as the integrity of the strand after placement forming the fibers will be of great importance for the membrane properties.

As mentioned in the introduction, the DIW technique allows controlled positioning of printed fibers. In other membrane fabrication techniques, such as electrospinning, positioning of the fibers is nearly random and lacks placement precision. Although it is possible to achieve pore size and fiber diameter in the range of a few nanometers, there is an inaccuracy within the structure, ranging from hundreds of nanometers to micrometers. **Figure 5** illustrates nanometer diameter electrospun fibers as well as presence of clusters, contributing to uneven structures and unpredicted pore sizes.

The direct ink writing setup in **Figure 4** allows us to move the substrate relative to the nozzle with sub-micrometer control. Various parameters could influence the size of the printed structures. To print 400 × 400 μm square structures, in **Figure 6(a,b)**, a nozzle with inner diameter of 6 μm and an outer diameter of 40 μm was used. In both structures, cellulose acetate (CA) was used as the ink, but the rheological properties of the ink differed. For **Figures 6(a)**, 5% CA with a molecular weight of 50 kDa was utilized, while **Figure 6(b)** employed 12.5% CA with the same molecular weight. All other printing parameters remained constant for both cases. The speed in the X and Y directions was set at 0.5 mm/s, and the ink flow remained constant at 3000 pl./s (calculated from syringe piston movement). The results illustrate that by using constant parameters and altering the ink concentration, significantly different outcomes can be achieved. In **Figure 6(a)**, controlling the ink flow through the nozzle was challenging, resulting in a printed line much larger than the nozzle's outer diameter. By increasing the amount of CA, properties such as the viscosity of the ink improved, which led to precise control over the printed fibers and resulted in the desired resolution, **Figure 6(b,c)**. In a structure with more than 10 layers printed, as shown in **Figure 6(c)**, the size of the printed fibers in the upper layers is close to the inner diameter of the nozzle, while the fibers size in the first layer is almost equal to the outer diameter of the nozzle. The larger ink spreading on the substrate in comparison with that on already printed material depends on several factors such as differences in wetting, solvent absorption, and solvent evaporation. The strand width is often controlled by the outer diameter of the nozzle in the first layer on

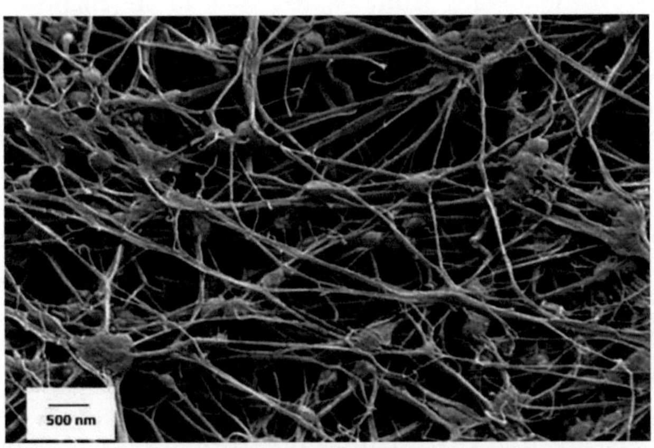

Figure 5.
Electrospun structure with uneven fibers and pore sizes.

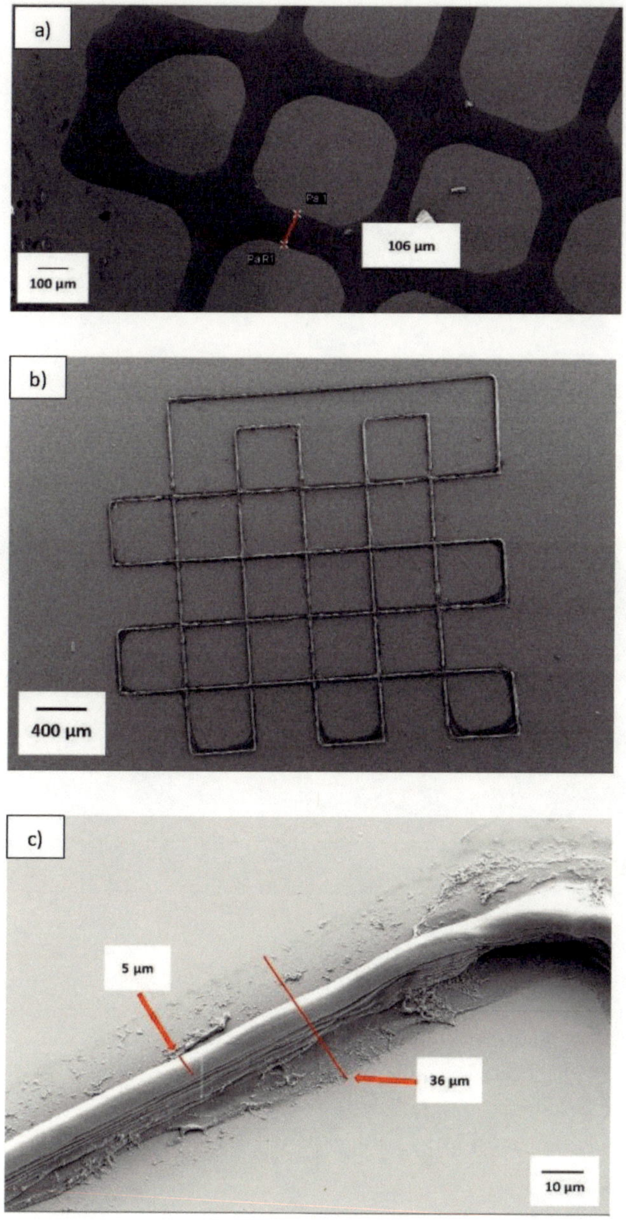

Figure 6.
Structures printed using inks containing 5% CA (a) and 12.5% CA (b, c). Sizes of the printed fiber in the first and upper layers are 36 µm and 5 µm, respectively (c).

non-absorbing substrates, while it is controlled by the inner diameter in the upper layers as observed in **Figure 6(c)**.

Optimizing parameters allows for the reduction of pore size as well as a reduction of fiber diameter giving larger porosity. To exemplify membrane structures, **Figure 7** demonstrates what occurs in the different layers during printing. An ink with 12.5% CA and a capillary with inner and outer nozzle diameters of 3 and 9 µm, respectively, were utilized. All other parameters were identical to those for **Figure 6(b)**. The

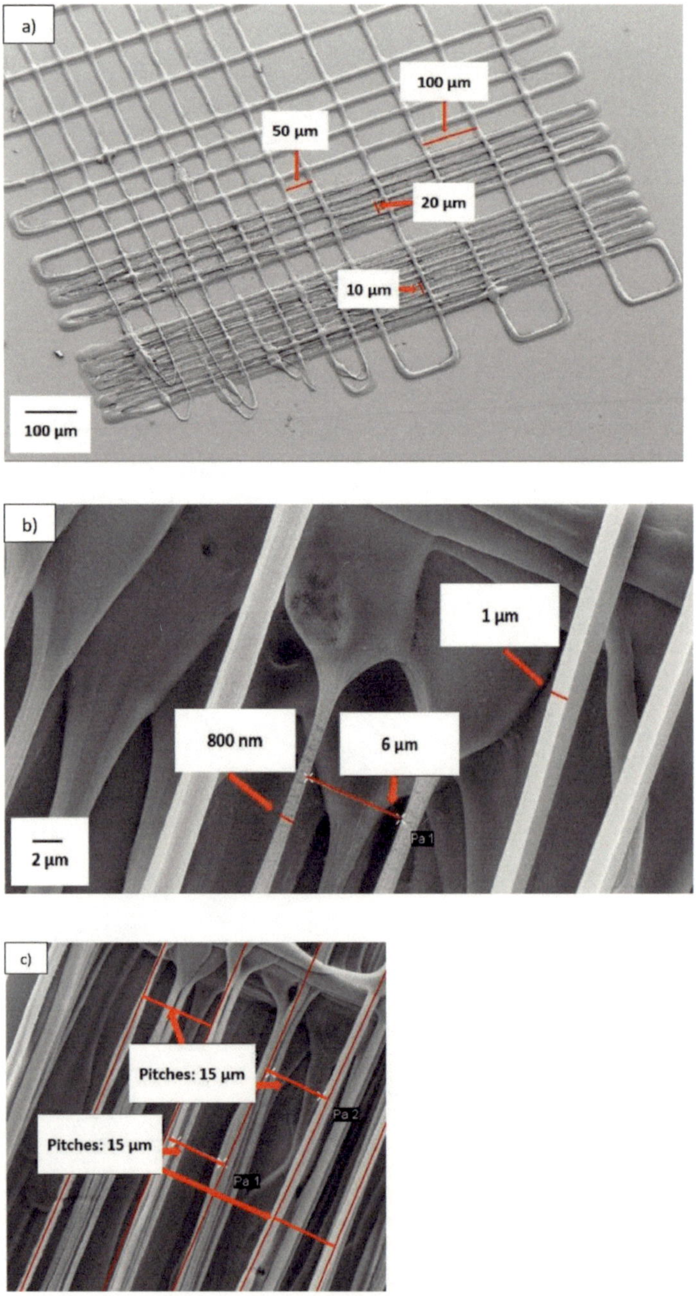

Figure 7.
Printed structures with 12.5%, (a) four layers structure with varying pore sizes, (b) inter-fiber distance of 6 µm when the pitch is 10 µm, (c) positioning of the printed strands with 15 µm pitch.

images demonstrate the capability to achieve controlled fiber spacings that would allow for varied pore sizes as designed for the structure. In the bottom layers close to the substrate, the ink typically spreads laterally, and this layer will typically be used to support the upper membrane layers. Moving from right to left in **Figure 7(a)**, the inter-fiber distance decreases from 100 to 50 µm, 30 µm, 15 µm, and finally 10 µm.

As can be noted, when the inter-fiber distance is reduced to 10 μm, fibers become attached to each other in these first layers. **Figure 7(b)** is the same structure with more than 15 layers, magnified to highlight the 6 μm inter-fiber distance between free-hanging printed fibers. The fibers in these upper layers are about 1 μm, allowing for large and controlled porosity. As exemplified in **Figure 7(c)**, where the pattern has a pitch of 15 μm in the designed structure, strands positioning is well controlled, which demonstrates placement control in the μm range in this DIW experiment. The deviations from the printing pattern are typically due to the capillary bending, resulting from a long and slender capillary to reduce risk of breakage.

3.5 Benefits and limitations

DIW has notable benefits for high-detail resolution 3D printing and, in particular, the ability to print micron-sized fibers with micrometer placement control. Membranes can therefore be made with well controlled pore size and

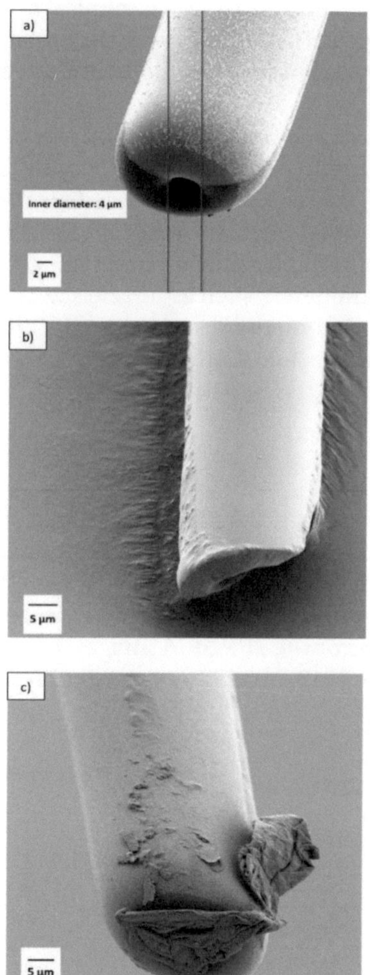

Figure 8.
(a) Nozzle with inner diameter of 4 μm, (b) nozzle with breakage, (c) nozzle with clogging.

size distribution. It is interesting to note that free-hanging strands can be printed rather large distances allowing advanced geometrical arrangements. The possibility of using a diverse range of materials makes it versatile and suitable for different purposes. However, these advantages come with some limitations, and for extrusion, it is necessary to prepare the material as an ink prior to printing. To achieve micrometer detail resolution, it is often necessary to prepare a capillary before each printing session, typically using a nozzle diameter within the range of a few micrometers. Typical challenges are frequent nozzle breakage and clogging, as shown in **Figure 8**. The ink therefore has to be well dispersed with no agglomerations larger than a fraction of the nozzle diameter. Furthermore, DIW has relatively low printing speeds, which makes it time-consuming to print large structures, and it is not yet optimal for industrial applications.

4. Conclusion

The chapter examines the 2PP and DIW techniques, both offering the potential for high-resolution 3D printing. 2PP technique gives the possibility of printing high-resolution structures; however, it demands photocurable resins, which today gives a rather limited material choice.

DIW appears to be a more versatile technique since there is a wide selection of printable materials. This versatility opens up new possibilities for applications in various industries, allowing for customization based on specific application requirements. The availability of a broad range of materials in DIW enhances its adaptability and makes it a compelling option for printing projects. Various parameters in DIW affect the final structure. Parameters such as nozzle size, ink rheology, printing speed, and ink flow have impact on the printing resolution. Balancing these factors is essential for achieving optimal resolution and desired printing outcome.

Acknowledgements

This work is conducted within the Additive Manufacturing for the Life Sciences Competence Center (AM4Life). The authors gratefully acknowledge financial support from Sweden's Innovation Agency VINNOVA (Grant no: 2019-00029). They acknowledge Myfab Uppsala for providing facilities and experimental support, in particular Victoria Sternhagen for some of the 2PP SEM images. Myfab is funded by the Swedish Research Council (Grant no: 2019-00207) as a national research infrastructure.

Conflict of interest

The authors declare no conflict of interest.

Author details

Farnaz Rezaei* and Stefan Johansson
Uppsala University, Uppsala, Sweden

*Address all correspondence to: farnaz.rezaei@angstrom.uu.se

References

[1] Truby RL, Lewis JA. Printing soft matter in three dimensions. Nature. 2016;**540**(7633):371-378. DOI: 10.1038/nature21003

[2] Liashenko I, Rosell-Llompart J, Cabot A. Ultrafast 3D printing with submicrometer features using electrostatic jet deflection. Nature Communications. 2020;**11**(1):1-9. DOI: 10.1038/s41467-020-14557-w

[3] Balčiūnas E et al. 3D printing hybrid organometallic polymer-based biomaterials via laser two-photon polymerization. Polymer International. 2019;**68**(11):1928-1940. DOI: 10.1002/pi.5909

[4] Fisch P, Holub M, Zenobi-Wong M. Improved accuracy and precision of bioprinting through progressive cavity pump-controlled extrusion. Biofabrication. 2021;**13**(1):015012. DOI: 10.1088/1758-5090/abc39b

[5] Placone JK, Engler AJ. Recent advances in extrusion-based 3D printing for biomedical applications. Advanced Healthcare Materials. 2018;**7**(8):1-11. DOI: 10.1002/adhm.201701161

[6] Huang H, Dean D. 3-D printed porous cellulose acetate tissue scaffolds for additive manufacturing. Additive Manufacturing. 2020;**31**:100927. DOI: 10.1016/j.addma.2019.100927

[7] Mohan D, Teong ZK, Bakir AN, Sajab MS, Kaco H. Extending cellulose-based polymers application in additive manufacturing technology: A review of recent approaches. Polymers (Basel). 2020;**12**(9):1876. DOI: 10.3390/POLYM12091876

[8] Hu Q et al. The influence of printing parameters on multi-material two-photon polymerisation based micro additive manufacturing. Additive Manufacturing. 2022;**51**:102575. DOI: 10.1016/j.addma.2021.102575

[9] Mahmood MA, Popescu AC. 3D printing at micro-level: Laser-induced forward transfer and two-photon polymerization. Polymers (Basel). 2021;**13**(13):1-27. DOI: 10.3390/polym13132034

[10] Huang Z, Deng Y. Two - photon polymerization nanolithography technology for fabrication of stimulus - responsive micro/nano - structures for biomedical applications 2 TPP PIs for biomedical applications. Nanotechnology Reviews. 2020;**9**:1118-1136

[11] Selimis A, Mironov V, Farsari M. Direct laser writing: Principles and materials for scaffold 3D printing. Microelectronic Engineering. 2015;**132**:83-89. DOI: 10.1016/j.mee.2014.10.001

[12] Faraji Rad Z, Prewett PD, Davies GJ. Parametric optimization of two-photon direct laser writing process for manufacturing polymeric microneedles. Additive Manufacturing. 2022;**56**:102953. DOI: 10.1016/j.addma.2022.102953

[13] Wu S, Serbin J, Gu M. Two-photon polymerisation for three-dimensional micro-fabrication. Journal of Photochemistry and Photobiology A: Chemistry. 2006;**181**(1):1-11. DOI: 10.1016/j.jphotochem.2006.03.004

[14] Slejko EA, Sesto Gorella N, Gasparini R, Scuor N, Seriani S. Tailoring 3D printed cellulose acetate properties produced via direct ink writing: Densification through over-extrusion and evaporation rate control.

Polymer Engineering and Science. 2023;**63**(11):3786-3797. DOI: 10.1002/pen.26484

[15] Rezaei F, Carlsson DO, Dahlstrom JH, Lindh J, Johansson S. Direct ink writing of high - resolution cellulose structures. Scientific Reports. 2023;**13**:22044. DOI: 10.1038/s41598-023-49128-8

[16] Heyman N, Heyman N. "High resolution 3d printing with cellulose acetate", microsystem technology division [thesis]. Sweden: Uppsala University; 2020

[17] Ligon SC, Liska R, Stampfl J, Gurr M, Mülhaupt R. Polymers for 3D printing and customized additive manufacturing. Chemical Reviews. 2017;**117**(15):10212-10290. DOI: 10.1021/acs.chemrev.7b00074

[18] Rodríguez K, Gatenholm P, Renneckar S. Electrospinning cellulosic nanofibers for biomedical applications: Structure and in vitro biocompatibility. Cellulose. 2012;**19**(5):1583-1598. DOI: 10.1007/s10570-012-9734-0